Nov 17th

Donald Stewart Strath Univ

WELDING METALLURGY OF STAINLESS AND HEAT-RESISTING STEELS

WELDING METALLURGY OF STAINLESS AND HEAT-RESISTING STEELS

R. CASTRO
Scientific Director, Ugine Aciers, France

J. J. de CADENET
Assistant Director, Ugine Aciers, France

Translated from the French by
R. C. JAIN
Welding Engineer
Babcock and Wilcox Co. Mount Vernon, Ind., USA

CAMBRIDGE UNIVERSITY PRESS

Published by the Syndics of the Cambridge University Press
Bentley House, 200 Euston Road, London NW1 2DB
American Branch: 32 East 57th Street, New York, N.Y.10022

First published as *Métallurgie du soudage des aciers inoxydables et résistant à chaud* by Dunod, Paris, 1968

Library of Congress Catalogue Card Number: 74–76582

ISBN: 0 521 20431 3

This English translation first published 1975

Printed in Great Britain
at the University Printing House, Cambridge
(Brooke Crutchley, University Printer)

CONTENTS

PREFACE

The basic requirements imposed on equipment used in various industries are being more frequently accompanied by demands which can be satisfied only by the use of noble materials such as stainless and heat-resisting chromium and chromium nickel steels.

The use of such equipment in service is often characterised by mechanical stresses applied at very low or very high temperature; it can also imply very severe risk of corrosion, or the requirement that rigorous working conditions be met without any chance of contamination. Moreover, in some cases, the chemical and mechanical requirements can be combined, particularly in cases of corrosion under static or dynamic stresses.

The diversity of applications has justified the development of a wide range of alloy steels, which may be put into service with complete confidence. Complicated fabrications are required involving various components which may be rolled, forged or cast. Experience has shown that conventional mechanical jointing methods – tacking, riveting, bolting – do not adequately meet the necessary requirements, and therefore welding is nearly always used, whatever difficulties may be encountered in its application.

With welded constructions and associated aspects, it is essential to know the welding metallurgy of the materials employed. Thus the present work of M. Castro and M. de Cadenet represents a most valuable contribution at the current time. In effect, the authors have taken the trouble to collate the essentials of their experience as metallurgists in the wide field of weldability.

By virtue of the methodical presentation and lucid style, the complex information gathered here will be easily utilised as a whole by readers, according to the different materials involved. Moreover, the authors of this book have been careful to base their arguments only on acquired experience or established reasoning. Thus their work offers advantages both in summarising existing knowledge, and in giving information for the solution, by transposition or analogy, of problems which may be met in future, as developments occur in materials or processes used in welded constructions. This constitutes a considerable asset for the present as well as the future, and M. Castro and M. de Cadenet, having had the courage to undertake the task, must be congratulated on it.

A. Leroy
Director of the Welding Institute, Paris

1

STAINLESS AND HEAT-RESISTING STEELS

1 General Characteristics

1.1 Stainless steels

The term 'stainless steel' is applied to a group of alloys which have the following main characteristics:

– Corrosion resistance in aqueous media in the presence of numerous aggressive organic and mineral agents; resistance to atmospheric corrosion is a particular case.

– Corrosion resistance at high temperature in gaseous or igneous environments.

Stainless steels are mainly alloys containing iron, chromium and often nickel, and it is to the presence of chromium that they owe their essential properties. Chromium is actually a less noble metal than iron in the electro-chemical series, and thus steels containing chromium may corrode, and especially oxidise, at a very rapid rate when they are not covered by a layer of chromium oxide; in this condition, they are termed 'active'. On the other hand, if such a layer can be formed, it is continuous, adherent, impermeable and virtually insoluble in numerous reagents; it protects the underlaying metal against the continuation of attack in certain corrosive media, particularly in oxidising environments, and the steel is said to be 'passive'.

This protective layer or 'passive film', is spontaneously formed by exposing bare metal to the air, or, more rapidly, by giving the metal a 'passivation treatment', such as immersion in an oxidising solution of 10–50 % nitric acid for a few minutes. When this layer has been removed, and if conditions are such as to oppose its spontaneous reformation (contact with a reducing acid, for example), stainless steel is little more resistant to corrosion than iron.

The favourable effect of chromium is apparent even when only a small amount of this element is present in the alloy and steels with only 5 % chromium are much more resistant to oxidation than ordinary steels. However, it has become customary to reserve the name 'stainless steel' to iron–chromium or iron–chromium–nickel alloys containing at least 10–12 % chromium. This percentage corresponds to the minimum generally required to obtain satisfactory resistance to atmospheric corrosion or spontaneous rusting.

The addition of nickel to iron–chromium alloys improves their corrosion resistance in certain media, mainly slightly oxidising and reducing environments. This element also has an appreciable effect upon the metallurgical structure, mechanical characteristics and weldability of stainless steels.

1.2 Steels for service at high temperature

The self-protection against continuous attack by oxygen, which characterises chromium steels at normal ambient temperature, is also found to give them good oxidation resistance at high temperature and special 'refractory' compositions have been developed. Moreover, a particular form of mechanical strength is required for these steels. The mechanical behaviour of metals at high temperature is appreciably different from their behaviour at normal ambient temperature, in that when the load is applied, a viscous, slow and irreversible deformation called 'creep' is superimposed on the reversible elastic deformation.

The creep resistance of iron–chromium-base steels is better than that of ordinary steels, and 13 % chromium steels, in particular, are widely used over a temperature range as high as 600 °C for certain types.

Nickel plays an important role in high temperature applications, because iron–chromium–nickel steels with an austenitic structure retain good creep resistance over a temperature range above that in which iron-chromium alloys are used.

1.3 Minor and other alloying elements

In addition to the main elements iron, chromium and nickel, stainless and heat-resisting steels contain smaller proportions of other elements, known as 'minor elements' and alloying elements. The most important of these are as follows:

– Minor elements such as carbon, silicon, manganese and nitrogen. These can arise from the basic materials used in steel manufacturing, or be deliberately added to facilitate its fabrication or modify certain properties.

– Alloying elements added in order to improve the properties of steels for specific applications. Apart from the minor elements mentioned immediately above, aluminium, molybdenum, copper, tungsten, titanium, niobium and cobalt are found in this category.

– Some elements are also present as impurities, and are generally undesirable. These are principally oxygen, phosphorus and sulphur, although the last named may be deliberately added in the steel to improve its machinability. Moreover, small amounts of other elements added to stainless steels to facilitate fabrication or change the material properties, can adversely affect the behaviour of these steels during welding.

1.4 Nickel-base alloys

To complete this brief description, mention must be made of nickel and sometimes cobalt-base alloys, containing high proportions of chromium, molybdenum and tungsten. Their relatively recent appearance has permitted the solution of problems of corrosion and strength at high temperature, for which standard stainless and heat-resisting steels have proved inadequate.

1.5 Types of stainless and heat-resisting steels

Apart from nickel or cobalt-base alloys, three main groups of stainless and heat-resisting steels can be distinguished according to their physico-chemical properties, i.e. according to the physical nature and chemical composition of the various phases formed in their structure at different temperatures. The importance of this classification lies in the fact that a knowledge of the physico-chemical properties of steels permits the prediction of their general behaviour, and particularly their weldability.

Alloying and minor elements have an appreciable effect upon these physico-chemical properties. However, to a first approximation, their effects can be related to those of the main elements, iron, chromium and nickel. On this basis, the three main groups of stainless and heat-resisting steels are described below by considering only these three principal elements.

Two categories are related to the iron–chromium system and the third one to the iron–chromium–nickel-system.

2 Steels based on the iron–chromium system

Figure 1.1 shows the equilibrium diagram of iron–chromium alloys. In that part of the diagram relating to the solid state, three regions are apparent:

– A portion limited by a loop known as the 'gamma loop', inside which the austenite phase is stable ('gamma phase'), having a face centred cubic crystalline structure.

– A portion situated outside a second loop enveloping the first one, where the body centred cubic ferrite phase ('alpha' or 'delta' phase) is stable from room temperature up to the melting point of the alloy. By analogy with iron and ordinary steels, ferrite formed on solidification is known as 'delta ferrite', while ferrite resulting from transformation of austenite during cooling is termed 'alpha ferrite'.

– A portion between the two loops, where both ferrite and austenite phases coexist, their respective proportions being dependent on the usual laws of equilibrium diagrams.

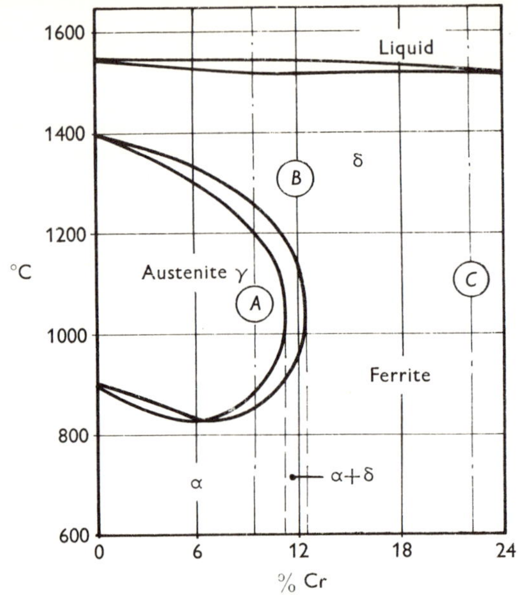

Fig. 1.1. Equilibrium diagram of iron–chromium alloys: *A*, martensitic alloys; *B*, semi-ferritic alloys; *C*, ferritic alloys.

From fig. 1.1, it is apparent that an increase in the chromium content has the effect of decreasing the temperature range over which austenite is stable. Up to 11 % chromium, the behaviour of the alloy resembles that of iron, although the temperatures at which ferrite transforms to austenite (Ac_3) and austenite transforms to ferrite (Ac_4) on heating become closer as the chromium content increases. Above about 13 % chromium, transformation does not occur on heating or cooling, and the alloy remains ferritic at all temperatures.

From this diagram, two categories of stainless steels can be identified:

Martensitic stainless steels having metallurgical characteristics similar to those of iron–chromium alloys containing less than 12 % chromium. These steels solidify as delta ferrite, but are completely transformed to austenite during cooling. When the cooling rate is sufficiently slow, transformation to alpha ferrite can take place. If the cooling is rapid, as is the case with welded joints, this transformation does not occur, and the austenite transforms to martensite, as with ordinary and low alloy steels.

Ferritic stainless steels which behave like iron–chromium alloys containing sufficient chromium to remain outside the external gamma loop. They are essentially ferritic over the whole temperature range corresponding to the solid state and consequently, they do not harden on quenching like martensitic steels.

Between these two categories, are martensitic–ferritic, or 'semi-ferritic', steels comparable to iron–chromium alloys found between the tangents vertical to the loops of the iron–chromium diagram. They consist partly of delta ferrite which remains unchanged at all temperatures in the solid state, and partly of delta ferrite which transforms to austenite and alpha ferrite or martensite, as indicated above in the case of martensitic steels.

The use of the iron–chromium diagram should be considered only as a convenient basis for distinguishing these different structural categories of stainless steels. The minor and alloying elements both act to form phases other than those present in the iron–chromium diagram, and particularly to displace the boundaries of the austenite and ferrite ranges. These effects will be considered further in the chapters dealing with the weldability of various types of stainless, refractory and creep-resistant steels. It is however necessary to mention specifically the importance of carbon in forming carbides, particularly with chromium, and in widening the range over which austenite is stable towards higher chromium content.

In practice, apart from the various elements referred to above, martensitic stainless steels usually contain 10–16 % chromium and up to about 1 % carbon or above. There are also other grades of martensitic steels containing up to 20 % chromium and less than 0·15 % carbon. With the addition of about 4 % nickel, these transform to austenite during cooling from very high temperatures, and subsequently may form martensite on continuous cooling down to room temperature.

Martensitic stainless steels are used either in the quenched and tempered condition, this treatment providing them with highest strength and ductility, or in the annealed state.

Ferritic steels are mainly represented by steels with 17 % chromium, containing 0·05–0·1 % carbon. As shown later on, this composition causes them to be 'semi-ferritic', comparable with iron–chromium alloys, traversing the two-phase range in fig. 1.1. In the form of plates or forgings, such steels are always utilised after annealing to give a structure of ferrite and carbides, but they can contain some martensite after rapid cooling, as in welding. Apart from the 17 % chromium grades, ferritic stainless steels used industrially include refractory 25 or 30 % chromium alloys, some of which never form austenite during heat treatment.

3 Austenitic steels based on the iron–chromium–nickel system

The introduction of increasing amounts of nickel into iron–chromium steels has the effect of widening the range over which austenite exists, and of increasing its stability at low temperatures.

This is illustrated in fig. 1.2, which shows schematically the influence of the nickel content upon an iron–chromium alloy, in which the chromium content is such that the alloy is purely ferritic in the absence of nickel (18 % chromium, for instance). With small additions of nickel, the steel

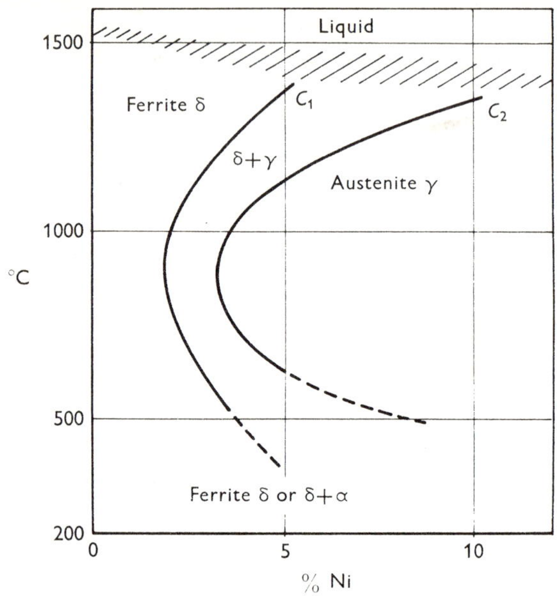

Fig. 1.2. Structure of a low carbon 18 % chromium steel with respect to nickel content and temperature. Below 5–6 % nickel, the austenite which is formed can transform to martensite.

continues to behave like an iron–chromium alloy. As the nickel content increases further, though remaining below a value C_1, the alloy still consists of delta ferrite at just below the solidus. However, at lower temperatures, it enters the two-phase austenite + ferrite range, and then changes completely into austenite if the nickel content is sufficiently high. Finally, with a nickel content above C_2, the alloy becomes completely austenitic just below the solidus.

At a given temperature, provided that this is high enough, the nickel content required to obtain a fully austenitic structure increases with increasing chromium content of the steel. For a given composition in which the nickel content is below C_2, there is a certain temperature at which some delta ferrite appears during heating, and above which the proportion of this ferrite increases as the temperature rises.

From the thermodynamic point of view, the austenite which is formed is unstable, and can actually be transformed, at least partly, into martensite, either by cooling to room temperature or below, or by plastic deformation at similar temperatures. In general, an increase in the quantity of all alloying elements makes this transformation more difficult, but the influence of nickel and carbon is particularly marked.

In practice, austenitic stainless steels generally contain more than 15 % chromium and a nickel content determined by the necessity to maintain a stable austenitic structure over the temperature range from 1100 °C to room temperature without the formation of martensite. The amount of nickel needed to ensure this stability will be considered more precisely in chapter 6, but it can be pointed out now that, for 15–16 % chromium, about 6–8 % nickel is required, allowing for the presence of the usual minor elements. Further, current steels with 18 % chromium and 10 % nickel and more highly alloyed steels are not very susceptible to martensite formation, even by moderate plastic deformation at room temperature.

The solubility of carbon in iron–chromium–nickel austenites decreases with decreasing temperature, and this phenomenon plays an important part in the metallurgy of such stainless steels. Current austenitic steels generally contain less than 0·15 % carbon, and this quantity is entirely soluble in austenite when the temperature is above 1000 °C. If these steels are cooled slowly, or maintained between 500 and 900 °C even for comparatively short periods, chromium carbides (which also contain other alloying elements) can precipitate, and can compromise certain service properties of these steels. This is the principal reason why austenitic stainless steels are employed after air or water quenching from 1000–1100 °C, because rapid cooling does not allow carbide precipitation. This solution treatment and quenching gives the steel a fully austenitic structure, with associated optimum properties of ductility and corrosion resistance.

The main classes of austenitic stainless and heat-resisting steels are given below, in order of decreasing tendency to form delta ferrite at high temperature:

- 17 % chromium and 7 % nickel steels.
- 18 % chromium and 10 % nickel.
- 25 % chromium and 20 % nickel.
- 18 % chromium and 38 % nickel.

They also contain the minor and alloying elements mentioned above, the influence of which will be studied in the chapter dealing with the welding of these steels.

2

THE WELDING OPERATION

1 Definition

A welding operation can be defined as a method of fabrication intended to give physical continuity between the pieces to be joined. In the great majority of practical cases, this continuity is realised by local fusion.

The quantity of molten metal appearing during the operation depends upon numerous geometrical and technological factors. When welding without added filler metal, this liquid metal results only from local fusion of the workpieces, while when filler metal is added, it results also from the complete fusion of the latter. The following fundamental characteristics are therefore always found in a welding operation:

- Rapid and local heating of the metal.
- Appearance of molten metal, at least some of which results from a partial fusion of the pieces to be assembled.
- Formation of a single molten pool.
- Rapid cooling of the whole.

All the metallurgical consequences of the operation are derived from these characteristics. The characteristics also permit differentiation between autogenous welding, and hard or soft brazing; in the latter joining processes, the molten metal is composed virtually only of filler metal, with a melting point much lower than that of the components being joined.

It must be noted that the term 'welding' is also applied to processes without added filler metal, in which molten metal is not apparent on a macroscopic scale, the metallurgical bonding being achieved by diffusion in the solid state. In this case, the pieces to be assembled must be in intimate contact, and the process must be capable of destroying the continuity of surface oxides or adsorption layers which may prevent diffusion. Pressure or ultrasonic welding processes come into this category, but they are as yet of little interest, particularly the latter, from an industrial point of view. On the other hand, the explosive welding process seems to offer interesting industrial prospects for the cladding of ordinary steel plates with stainless steel. In this process, oxides and adsorption layers are broken up, and the cladding operation performed by a shock wave produced by an explosion. The high pressures produced by this shock wave cause either local fusion or solid state diffusion which ensure a continuous metallurgical bond between the workpieces to be assembled.

These processes will not be considered further, attention being concentrated on the following categories, which correspond to the classical definition and cover most stainless steel welding applications:

– 'Fusion' processes in which the workpieces to be assembled are simply placed in contact with each other and joined by a liquid molten pool produced by a heating source external to the joint.

– Autogenous processes in which the workpieces to be assembled are pressed together with the formation of small amounts of liquid metal.

2 Heat distribution in a fusion welding operation

An electric-arc weld will be used as an example, as shown in fig. 2.1. The arc is struck between a metallic plate, whose surface coincides with the plane xOy of the figure, and an electrode moving with constant speed parallel to Oy. By plotting the temperature of the different points of the plate along the axis Ot at the instant the arc passes the point O, a surface is obtained representing the instantaneous thermal state of the metal. The thermal gradients are very steep in the immediate vicinity of the arc, and become progressively less steep as the distance from the arc increases. The surface travels along Oy with the same speed as the arc. Thus, a section through this surface at a plane $x = x_1$ represents the change in temperature with respect to time, experienced at a point situated on the plate at a distance x_1 from the arc trajectory, the time axis being shown by Oy.

The thermal gradients in space or time are steeper as the thermal conductivity of the metal is lower. The table below shows that the thermal conductivities of stainless steels are lower than those of ordinary steels, particularly in the case of austenitic steels. This is the reason why heating due to welding is generally much more localised, and cooling after welding is slower with stainless steels than with ordinary steels. This can have unfavourable consequences, as will be considered in §3.

This type of heat distribution is valid whatever the heating source, and has been the subject of various theoretical studies [1]. Direct measurement [2] has also been used to determine the thermal cycles experienced at different points of an 18 % Cr/10 % Ni/Nb stainless steel plate, on which a fusion run is made by an electric arc. Representative results are shown in fig. 2.2.

3 Mechanical consequences of a welding operation

Heating during the welding operation leads to an expansion of the pieces being joined. As areas unaffected by heat cannot generally move freely, the edges of the joint approach each other, so that the liquid metal occupies a smaller volume than that existing at room temperature. During cooling,

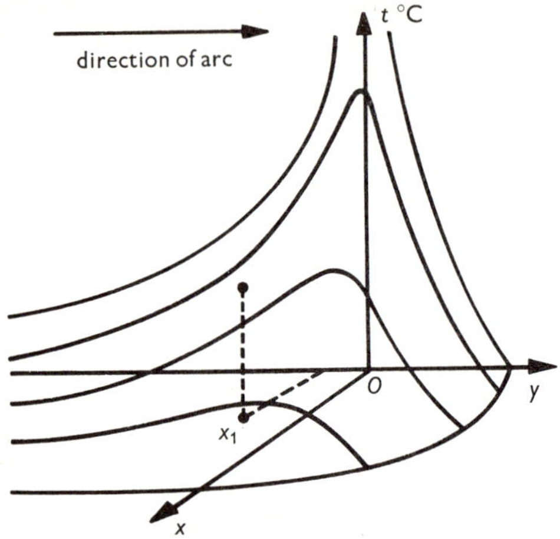

Fig. 2.1. Heat distribution in a fusion welding operation.

Table 2.1. *Average thermal characteristics of ordinary steel and various stainless steels at room temperature*

	Thermal conductivity (c.g.s.)	Coefficient of expansion (per °C)	Thermal capacity (c.g.s.)
Ordinary steel	0·140	12×10^{-6}	0·12
Martensitic steel with 13 % Cr	0·065	10×10^{-6}	0·12
Ferritic steel with 17 % Cr	0·060	11×10^{-6}	0·12
Austenitic steels			
18 % Cr/10 % Ni	0·035	17×10^{-6}	0·12
25 % Cr/20 % Ni	0·030	$14{\cdot}5\times10^{-6}$	0·12
18 % Cr/38 % Ni	0·030	14×10^{-6}	0·12

the assembly tries to regain its original size and this contraction, together with shrinkage of the molten metal during solidification, subject the weld to a multiaxial stress system. Analysis of the effects of the thermal characteristics of the steel upon the level of these stresses is a complex problem, which will not be considered here. It must, however, be noted that austenitic stainless steels behave quite differently from ordinary steels as a result of their higher coefficients of expansion, their lower conductivity (which increases the thermal gradients) and their particular elastic and plastic properties at high temperatures. In respect of thermal stress alone, martensitic and ferritic steels appear comparable to low alloy steels, except for their thermal conductivity.

The existence of these stresses can have the two following consequences:

– The fabrication is stressed during cooling, even though its mechanical

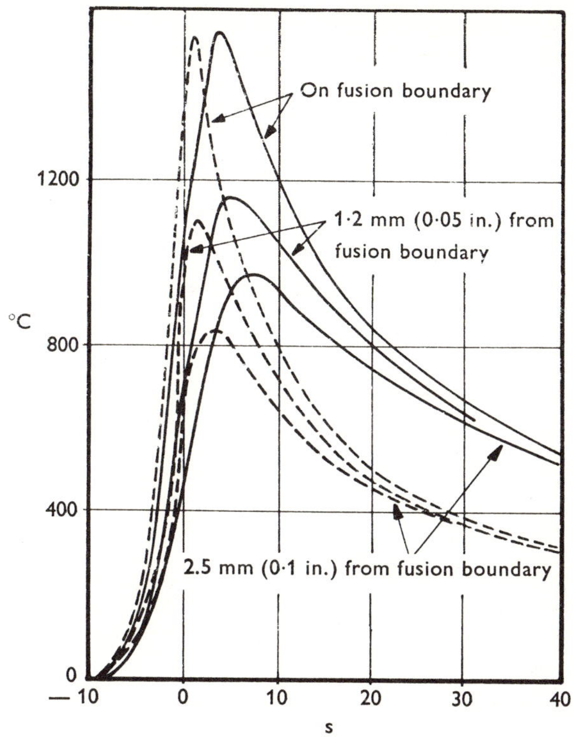

Fig. 2.2. Thermal cycles due to a welding arc on an 18 % Cr/10 % Ni/Nb steel plate, 38 mm (1·5 in.) thick, for two arc energies. ——2·8 kJ/mm (70 kJ/in.); ----- 1·4 kJ/mm (35 kJ/in.).

strength may be reduced by the high temperature, or by the formation of brittle structures. This can lead to cracking in the weld area. The metallurgical factors associated with cracking will be considered in chapters 4 and 6, dealing respectively with the welding of martensitic and austenitic stainless steels. However, it should be appreciated that the fundamental cause of cracking lies in the presence of shrinkage stresses. Thus, conventional technological measures aimed at reducing the stress level can be applied to stainless steels at least as effectively as to ordinary steels, provided that these measures are compatible with the particular metallurgy of the steel under consideration.

– After cooling, the weld is subject to internal stresses. These stresses result from the shrinkage, thermal cycles and thermal heterogeneities created by welding; they can be partly relaxed during cooling of the weld, particularly if preheat or post-heat is applied. This relaxation is caused by plastic deformation, and is easier as the elastic limit is reduced at a given temperature.

Generally, stainless steels, and especially austenitic steels, have high mechanical characteristics at elevated temperature, and this makes the relaxation of welding stresses difficult. Welds made in these steels can be subject to very severe residual stresses, in addition to anticipated service stresses, and when welding thick plates, it may be advisable to specify a 'stress relieving' treatment to allow their relaxation.

Finally, the welding thermal cycle gives rise to more severe deformation with thin gauge stainless steel than in the case of ordinary steels.

4 Microstructural and compositional effects

The properties of the steel which the user puts into service are defined both by its chemical composition and by the heat treatment to which it has been subjected. A welding operation sometimes affects the chemical composition of the steel, when filler metal is added to the joints, and always alters the effect of the heat treatment.

Welding is a 'micrometallurgical' operation which causes the formation of two particular zones, shown schematically in fig. 2.3:

- Molten metal cooled very quickly from the liquid state.
- Unmelted metal in which the rapid and non-uniform thermal cycle has caused structural changes, which are often irreversible. This region is called the 'heat-affected zone' (HAZ).

The structures formed in these two regions depend on the type of stainless steel. However, they have some general characteristics which are described below.

4.1 The weld metal

The weld metal possesses the dendritic solidification structure characteristic of cast metals, often with a columnar structure orientated perpendicularly to the isotherms as observed in the solidification of ingots. This is illustrated in fig. 2.4.

This heterogeneous orientation is associated with chemical inhomogeneity resulting from the solidification interval of the steel, and from the relatively slow diffusion in the solid state. The mechanism of this segregation is shown in fig. 2.5.

During solidification, the liquid and the liquid–solid interface follow the equilibrium curves approximately. However, the composition of the dendrite cores describes a curve which departs further from the equilibrium curve as the cooling rate is increased. The development of the average composition of the solid can be represented by an intermediate curve. This has the two following consequences:

- The solidification interval is increased compared to equilibrium conditions.

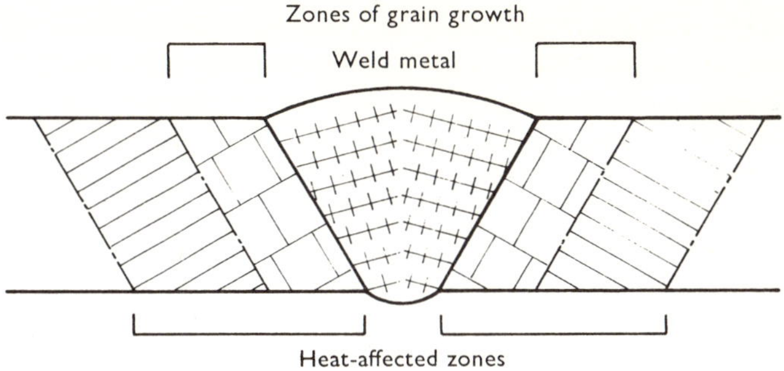

Fig. 2.3. Weld metal and heat-affected zones.

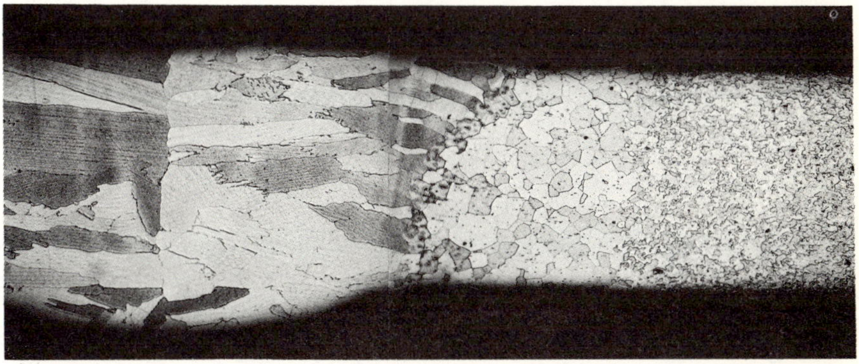

Fig. 2.4. Structure of a TIG weld in a 17 % chromium ferritic steel, containing an addition of niobium (× 35).

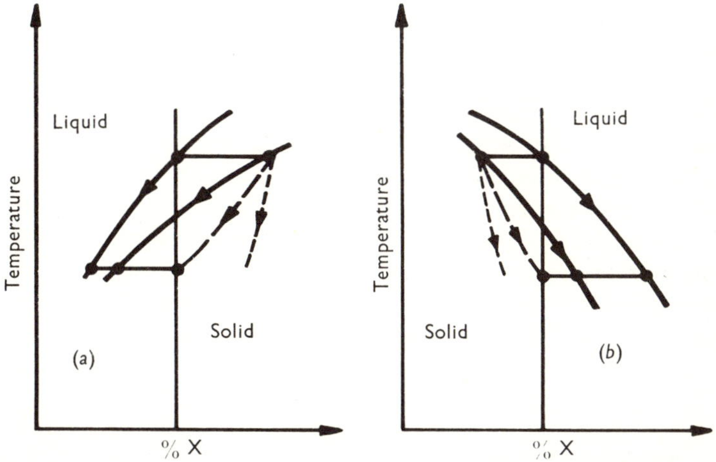

Fig. 2.5. Segregation during solidification: —— composition at solid and liquid equilibrium; ----- composition of the dendrite cores; ——— average composition of the solid.

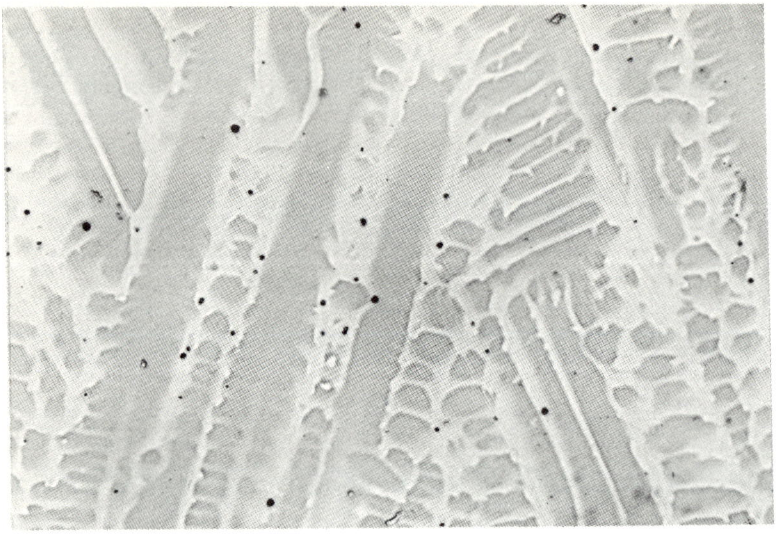

Fig. 2.6. Chemical heterogeneity in the weld metal of a joint in an 18 % Cr/13 % Ni austenitic steel (× 450).

– Depending on whether the equilibrium diagram corresponds to fig. 2.5(*a*) or (*b*), the interdendritic regions are impoverished or enriched in element X, as compared to the average metal composition, whereas the dendrite axes are enriched or impoverished.

This also occurs with an ingot, but the slow cooling together with the thermal and mechanical treatments leading to the final product, tend to reduce this inhomogeneity. A weld is often put into service in the as-welded condition, and this heterogeneity, which can even cause the formation of physico-chemical constituents that would not exist in a forged and heat-treated metal of identical composition, is likely to have an influence on the applied properties.

This segregation can be shown by chemical attack of a polished surface, as the differences in chemical composition lead to different dissolution rates in the etching reagent (fig. 2.6). Further, segregation can be quantitatively studied by electron-probe microanalysis, as considered in chapter 6 dealing with the austenitic stainless steels, in which segregation plays an important role.

These two types of heterogeneity are more pronounced with smooth, steady solidification. They are restricted by agitation of the liquid pool, as occurs when welding with covered electrodes.

4.2 The heat-affected zone

In the unmelted metal close to the fusion boundary, the rapid thermal cycle can also cause structural transformations, if the maximum temperature attained is sufficiently high to nullify the effect of previous heat treatment. Moreover, the material immediately adjacent to the weld metal, which is heated to a very high temperature, is subjected to grain growth (fig. 2.4) and sometimes to local fusion of the grain boundaries causing the formation of constituents having a particular structure.

Finally, the rapid cooling which follows welding is analogous to quenching from different temperatures, and can leave non-equilibrium structures in the weld and HAZ at room temperature. The nature of these structures depends upon the type of steel being considered.

5 Conclusions

It is apparent that the mechanical and structural state of a weld is quite different from the initial state of the base metal. The ability of the weld to withstand shrinkage stresses without failure, and to satisfy service requirements, depends upon the chemical heterogeneities and the nature of constituents produced by the welding thermal cycle. The three main categories of stainless and heat-resisting steels show markedly different structures after fast cooling from very high temperatures, and will be considered separately.

References

[1] A. Portevin and D. Séférian. 11e Congrès international de l'acétylène et de la soudure autogène, Rome, 1934.

[2] E. F. Nippes, H. Wawrousek and W. L. Fleishmann. *Weld. J. Res. Sup.* **34** (1955) 169 s.

3

APPLICATION OF THE FUSION WELDING PROCESSES TO STAINLESS AND HEAT-RESISTING STEELS

All fusion processes used for ordinary and low alloy steels are applicable to stainless and heat-resisting steels, with some precautions which depend on the type of steel employed. These will be briefly reviewed, after considering the importance of protecting the weld metal against gases arising from the atmosphere or welding consumables.

1 Necessity for weld metal shielding

The action of oxygen, nitrogen and hydrogen on the molten metal can affect the physico-chemical properties of a weld, or create physical defects which seriously impair its mechanical strength.

1.1 Physico-chemical effects

Oxygen. Certain elements of the weld metal are readily oxidised at elevated temperature, and their preferential oxidation during welding can lead to a change in the properties of the joint. This is the case with chromium, the removal of which reduces the corrosion resistance of the weld deposit; titanium also oxidises very easily, and the consequence of this will be considered with particular reference to austenitic steels.

Moreover, the oxides formed under these conditions often have a very high melting point, and their presence can prevent the liquid metal from wetting the base metal or the underlying weld metal. This results in lack of fusion defects, which are detrimental to the mechanical strength of the weld, and can cause local corrosion in humid or aqueous media, by a phenomenon known as differential aeration.

Nitrogen. Nitrogen is highly soluble in chromium and chromium–nickel steels. Its effect on the structure of these steels is similar to that of carbon, and the consequences which this can have on the welding of the different types of steels will be considered below.

It can be pointed out here that this gas may be deliberately introduced in the arc for welding certain types of martensitic and austenitic steels, but that it should be avoided in the case of ferritic steels.

In general, the nitrogen content normally present in weld deposits containing 13 % chromium and above, varies between 0·03 and 0·07 %.

Hydrogen. Hydrogen has no marked physico-chemical effect on austenitic steel welds. This allows these steels to be welded under shielding atmospheres containing significant amounts of this gas, as discussed below.

Hydrogen must be avoided with ferritic steels, and particularly martensitic steels, in which it causes the formation of porosity or of a particular type of cracking.

1.2 Formation of porosity

In general, the solubility of oxygen and hydrogen in stainless steel is much higher in the liquid state than in the solid state. Consequently, rapid gas evolution can occur during solidification, giving rise to spattering of liquid metal, or to porosity when the gaseous bubbles remain entrapped in the solidified metal. This tendency to form pores due to oxygen and hydrogen seems to increase as the nickel content of the steel increases, and becomes quite pronounced in nickel-base alloys.

The nature of the gases inside pores is not well known. Generally, the pores have bright inside surfaces, which indicates that oxygen is not involved in their formation. However, oxygen can act indirectly by forming gaseous compounds with carbon when the steel contains 0·1 % or more of this element.

The high solid solubility of nitrogen in chromium steels suggests that this gas does not play a part in the formation of blowholes, at least as long as the nickel content does not exceed 30 %. It seems, however, that nitrogen plays an important role in nickel-base alloys, as shown in chapter 7.

Hydrogen is, without doubt, mainly responsible for the formation of pores in stainless steel welds. It can have a direct effect as there is a considerable difference between the solubilities of this gas in the liquid and solid phases. In iron, the limit of hydrogen solubility is reduced by a factor of 3 or 4 during solidification, and it is presumably this simple phenomenon that causes porosity in ferritic steel welds made under argon containing about 10 % hydrogen (fig. 3.1).

Figure 3.2 shows that hydrogen is more soluble in iron in the austenite phase than in alpha or delta ferrite. It can be assumed that, in stainless steels, the risk of porosity is reduced when the alloy solidifies with a high proportion of austenite, as is the case with austenitic steels containing at least 10 % nickel. Indeed, it has been found that these steels can be welded without porosity under argon containing up to 10 % hydrogen. This proportion of hydrogen probably constitutes the maximum that should be used, particularly when the steel composition is such that the solidification structure is mainly ferritic; it can be too high in certain cases of welding 18 % Cr/10 % Ni type austenitic steels, notably when oxides are present

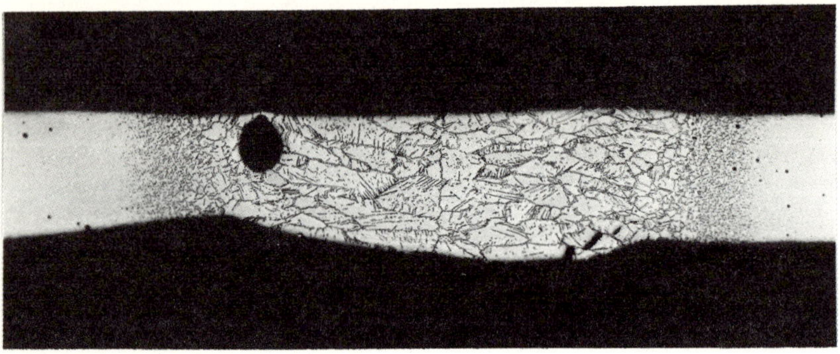

Fig. 3.1. Pore in a 17 % chromium steel weld carried out under argon with 10 % hydrogen (× 15).

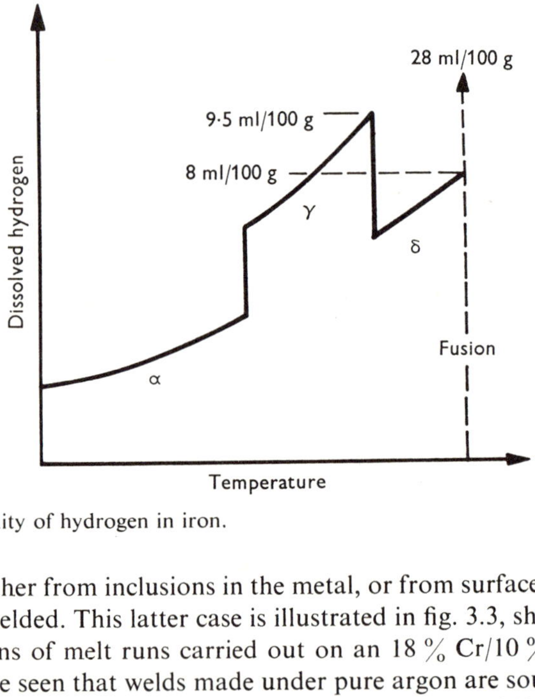

Fig. 3.2. Solubility of hydrogen in iron.

originating either from inclusions in the metal, or from surface scale on the pieces to be welded. This latter case is illustrated in fig. 3.3, showing longitudinal sections of melt runs carried out on an 18 % Cr/10 % Ni/Ti steel plate. It can be seen that welds made under pure argon are sound, whether the plate is covered with scale or not; under argon with 10 % hydrogen, the weld on a ground plate is sound, but that carried out on a scaled plate contains very large pores.

A similar but less pronounced phenomenon can occur when the metal contains many oxide inclusions. The metal composition, the quantity of oxygen in solution in the liquid metal, and the presence of solid oxides [1] have a marked influence upon the amount of hydrogen likely to be dis-

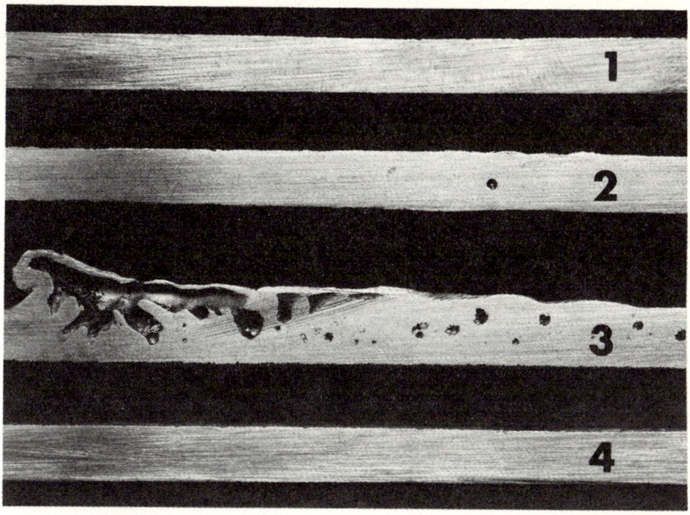

Fig. 3.3. Arc welds on 18 % Cr/10 % Ni/Ti austenitic steels (× 1·5): 1. pure argon, ground plate; 2. argon with 10 % hydrogen, ground plate; 3. argon with 10 % hydrogen, scaled plate; 4. pure argon, scaled plate.

solved in the liquid metal, and consequently liberated on solidification. However, in view of the rapidity of a welding operation, caution is necessary in using experimental results obtained under conditions more related to the equilibrium state. It is also possible that some oxides present in stainless steels or on the surface may be reduced by hydrogen at the arc temperature to form gaseous compounds.

Hydrogen may also be introduced in weld metals in the form of moisture resulting from electrode coatings, or fluxes, or from the presence of humidity in shielding gases (condensation, leakage in cooling circuits, etc.). It is generally found that, under these conditions, the incidence of porosity is higher in chromium steels than in ordinary steels. Humidity may also have a highly detrimental effect in nickel-base alloys.

An indirect effect of hydrogen can further be encountered in the presence of carbon, sulphur and perhaps selenium, since it can react with these elements to form gaseous compounds. Such reactions are responsible for causing porosity during the welding of free machining stainless steels containing sulphur or selenium in the presence of hydrogen or humidity.

The formation of gas bubbles by chemical reaction and their elimination by decantation depend upon the temperature, viscosity and surface tension of the liquid metal, and upon the time. Consequently, all physical and technological conditions influencing these factors, and particularly those which control the liquid metal temperature and its solidification rate, play a substantial role in the formation of porosity.

2 Oxy-acetylene welding [2]

In this well-known process, the combustion of an oxygen–acetylene mixture provides the heating source as well as the shielding atmosphere.

With a neutral flame, the combustion essentially produces carbon monoxide and hydrogen, which then burn in the outer cone in contact with atmospheric oxygen. This flame is especially suitable for ordinary steels, as the atmosphere corresponds more or less to the carbon–oxygen equilibrium in the molten metal. Although this advantage does not apply in the case of stainless steels, particularly when they have low or very low carbon contents, a neutral flame is recommended, combined with the use of a flux.

Nevertheless, it is possible to use a slightly oxidising flame for low carbon steels in which there is a danger of carburisation, or a slightly carburising flame when depositing a multipass overlay of hard 13 % chromium martensitic stainless steel.

The oxy-acetylene process is being largely replaced by the electric-arc methods since it has a number of disadvantages in the welding of stainless steels:

– The oxidising flame forms refractory oxides which may be difficult to remove because of the moderate temperature of the heating source. It is therefore essential that a flux is used.

– The thermal efficiency is low. Only a small amount of the heat produced is used to melt the metal; the rest heats a large area of base metal, which causes severe deformation and may be detrimental to the corrosion resistance of austenitic steels.

– The risk of carburisation involved in oxy-acetylene welding always leaves some doubt regarding the ability of austenitic welds in very low carbon material to resist severe corrosion conditions.

3 Electric-arc processes [3]

3.1 Distribution of the energy consumed

The heat necessary for welding is furnished by an electric arc struck between an electrode and the workpiece to be welded. The total expenditure of energy in the arc is, of course, equal to the product of the voltage drop existing between the electrode and the workpiece, and of the current flowing across the arc.

Three regions can be distinguished in terms of the distribution of the voltage drop (fig. 3.4):

– A cathode zone with a thickness of the order of microns, where electrons are emitted, their movement contributing about 99 % of the current flowing in the arc.

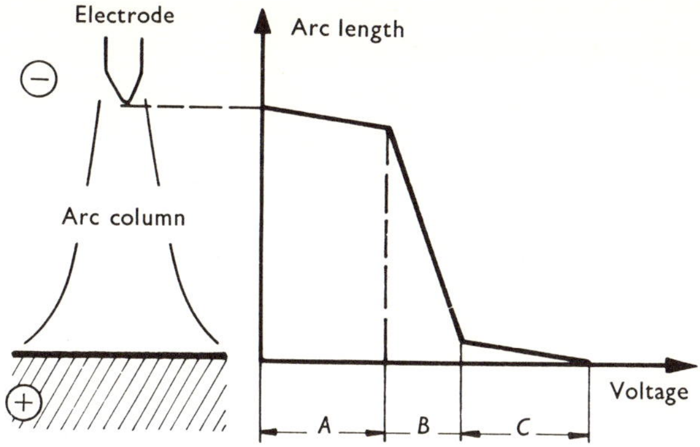

Fig. 3.4. Voltage drop distribution in an electric arc: *A*, cathode drop; *B*, drop in arc column; *C*, anode drop.

– An anode zone of similar thickness, where there is an emission of ions also participating in the arc current.

– An arc column, occupied by a 'thermal plasma' composed of ions, electrons and atoms of the shielding gas.

The dissipation of energy in the plasma occurs by loss of electrons or ions and by radiation. This dissipation is low, and the voltage drop in the arc column represents less than one third of the total voltage drop. Most of the voltage drop is localised in the thin cathode and anode regions, so that arc processes have a very high efficiency.

The cathode region plays a major role in the stability of the arc, as it is in this region that the electrons are produced. These electrons are produced by thermionic emission and by emission due to an electric field having a magnitude of about 10^6 V cm^{-1}. These two mechanisms absorb a very large part of the energy consumed in the cathode region.

The electrons reaching the anode region are accelerated by the electric field prevailing in this zone. At the anode, they produce a part of the ions necessary to maintain the arc column (about 1 ion for 1000 electrons). Production of ions absorbs little energy, and most of the energy released at the anode appears in the form of thermal energy, owing to the bombardment of electrons. In consequence, heat liberation at the anode is considerably greater than that at the cathode. In addition, it must be pointed out that a plasma current exists, directed from the electrode towards the workpiece, regardless of the polarity selected (see §3.2). This current also contributes in heating the workpiece to be welded.

3.2 Metal transfer

In certain cases, the electrode is consumable and melts to provide the filler metal required for the welding operation, as a result of the energy dissipated as heat at the cathode or anode. Drops of liquid metal are detached from the electrode and fall (or rise) towards the workpiece, whatever polarity is selected.

The transfer of metal in the arc can be explained in the following way [4].

The liquid metal drop is detached from the electrode by an electromagnetic effect known as 'pinching', which tends to attract the peripheral regions of a conductor towards its axis when a current is flowing. This detachment produces an impulse which projects the liquid drop towards the workpiece to be welded. The plasma is also a conductor and is equally subjected to the pinch effect. It can be shown that, taking into account the shape taken by the plasma as shown schematically in fig. 3.4, this pinching effect is associated with a magnetic pressure gradient which causes a stream of plasma from the electrode to the workpiece, regardless of the polarity. The drops of liquid metal are subjected to this pressure gradient, and are thus transferred across the arc. Transfer is also caused by physical entrainment of the drop in the plasma stream.

It must be added that this plasma current is facilitated by suitable control of the welding conditions with the gas-shielded welding processes.

4 Manual metal-arc welding

The principle of this process is shown in fig. 3.5. The consumable electrode is composed of a metallic core acting as filler metal and carrying the current, and covered with a coating consisting mainly of non-metallic ingredients.

The covering melts at the same time as the core, and plays an important part in the physics and metallurgy of the process.

4.1 Part played by the coating

Protection of the molten metal. The coating forms a slag which covers the metal during solidification and protects it against the atmosphere. During melting, it forms a sunken cone shape as represented in fig. 3.5, and restricts the entrance of air into the plasma, provided that a short arc is maintained. The coating probably also covers the melting tip of the metallic core with a thin liquid film.

In addition, the covering provides gases which form the arc plasma. In contrast to the case with ordinary steels, cellulosic electrodes are not used for stainless steel welding, and the gases liberated by the heat of the arc arise essentially from the decomposition of alkaline earth and alkaline carbonates, which are always found in considerable quantity in the coverings of stainless steel electrodes. This atmosphere also protects the molten metal against oxidation.

However, this protection is not entirely effective, and the melting of a

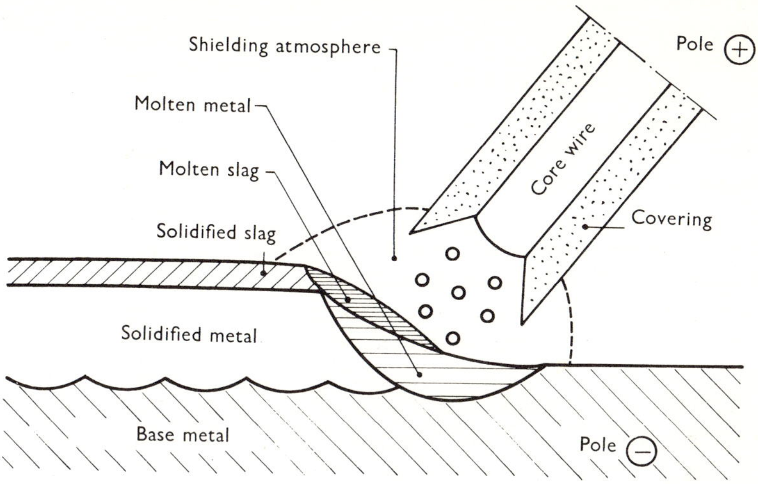

Fig. 3.5. Principle of the manual metal-arc process.

covered electrode is accompanied by a loss of easily oxidisable elements. This loss is partly due to oxidation by the air which penetrates the plasma, but it can also result from reactions occurring between reducing elements contained in the core, and the constituents of the coating. For instance, this happens with chromium, which can react with carbonates to form complex oxides containing chromium and calcium. It is also the case with titanium. This is used as an alloying element in some austenitic stainless steels, but it disappears almost completely during manual metal-arc welding by oxidation due to air and by reduction of silicates present in the coating. This results in an increase in the silicon content in the weld metal, which may have undesirable consequences. Further, certain elements with relatively low boiling points can volatilise in the arc when the metal transfer is sufficiently slow.

The combined effect of these different phenomena leads, for example, to a loss of 1–3 % chromium in absolute value during the melting of an electrode with a core of 18 % Cr/10 % Ni steel, covered with a chromium-free coating. To compensate these losses, the manufacturers of covered electrodes either use stainless steel wires having higher contents of reactive elements than are required in the weld, or incorporate metallic powders into the coatings.

Finally, coatings can contain moisture causing porosity and increasing the cracking susceptibility of martensitic steels. The best protection of the molten pool is obtained only with electrodes baked in order to expel humidity.

Arc stabilisation. The volt–ampere characteristic of an electric arc is shown in fig. 3.6. At low currents, the plasma temperature (in the sense of

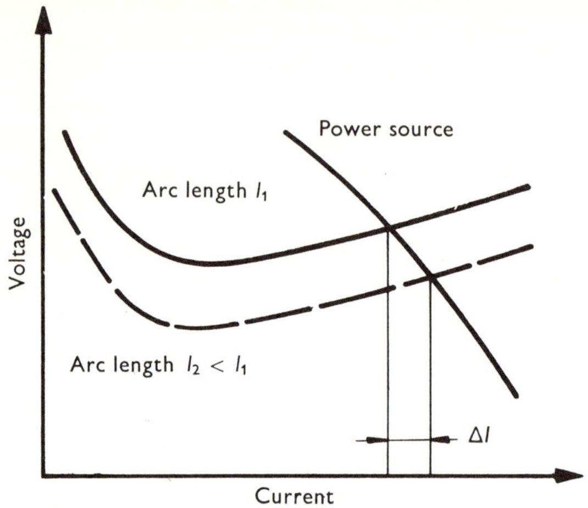

Fig. 3.6. Arc and power source characteristics.

the mobility of the constituent particles) is low, and consequently the degree of ionisation is low. An increase in current arises mainly through an increase in the degree of ionisation, and hence through a reduction in the voltage drop. At high currents, the number of ionised particles is little further increased, but their average speed (and so the current) increases simultaneously with the potential difference.

All factors facilitating ionisation of the plasma have a favourable effect on arc stability, because for given power source characteristics, they tend to reduce the threshold current below which the system becomes unstable.

At constant current, the potential drop increases with increasing arc length. Since coated electrodes are used manually, the unavoidable variations in arc length caused by movement of the operator's hand lead to current variations (ΔI in fig. 3.6), which depend upon the electrical characteristic of the power source. This affects the uniformity of the weld, because the penetration and melting rate of the electrode depend mainly upon the current. These variations are reduced when the power source characteristic is steep in the current range over which the electrode functions. The situation is further improved when the energy required to produce ions and electrons is low.

The coatings therefore contain mineral compounds with a low ionisation potential, usually silicates and alkaline carbonates, and sometimes oxides with a high emissive power, the most common being titanium dioxide, found naturally as rutile. The presence of these compounds also facilitates arc striking, and enables certain electrodes to function with alternating current.

Role of flux and slag. The fluxing compounds present in coatings for ordinary steels are generally insufficient in the case of stainless steels, because of the refractory nature of the oxides formed on the latter alloys. To avoid the formation of lack of fusion defects, coatings for stainless steels contain energetic fluxing agents such as alkaline earth and alkaline fluorides, or cryolite. However, these compounds have an unfavourable effect on the arc stability.

While melting, the coating forms a slag which fulfills the usual functions of metallurgical slags. Deoxidation is particularly important in the case of welding, and is ensured by the presence of silicon, manganese or ferro-manganese. The viscosity of the slag is significant for the following reasons:

- The metal–slag chemical reactions are facilitated by a high slag fluidity.
- The welding bead is smoother and of better appearance with a fluid slag.
- The convexity of the bead, and the ability of the electrode to be used in positions other than downhand are influenced by the metal–slag surface tension, and the slag viscosity.

Control of the weld metal composition. It has already been pointed out that the electrode coating may be used to compensate for loss of easily oxidisable elements. Thus coverings for stainless steels generally contain an important proportion of chromium or ferro-chromium in the form of powder.

This method is not limited only to compensation of losses during fusion. In addition, it permits the manufacture of electrodes giving deposits of different composition, or containing various additional elements, while always using the same core which only contains the main elements – chromium and possibly nickel. Thus, 19 % Cr/10 % Ni/2·5 % Mo electrodes are often fabricated with 19 % Cr/10 % Ni wire, molybdenum powder being incorporated in the coating.

Similar types of electrodes are also manufactured with a dead mild steel core wire, all the alloying elements being contained in the coating. Owing to the low electrical resistivity of the core, these electrodes can be employed with high currents, and they therefore have the advantage of a higher deposition rate than that of electrodes with a stainless steel core of similar diameter. Unfortunately, the metal deposited by these electrodes can be of non-uniform chemical composition on the microscopic and sometimes macroscopic scale when the covering is damaged. Moreover, the molten metal spatters which always occur during the fusion of a covered electrode can deposit adherent drops of ordinary steel on the base metal. Both these factors can reduce the corrosion resistance of the weld, and it is advisable to use this type of electrode only when corrosion conditions are not severe.

The possibility of effecting compositional control through the coating imparts considerable flexibility to the process as applied to stainless steels, particularly since it permits the ready manufacture of small series of electrodes meeting specialised and limited applications.

4.2 Principal types of coatings for stainless steels

Five main types of coating are, or have been, used for ordinary or low alloy steels [5]:

– Oxidising coatings giving iron oxide slags with or without manganese oxide.

– Cellulose coatings, which are rich in organic material and give a rather high hydrogen content in the deposited metal.

– Acid coatings giving slags based on oxides of iron, manganese and silicon.

– Rutile coatings giving slags with a high proportion of titanium oxide.

– Basic coatings containing a high proportion of alkaline earth carbonates, with the slags rich in lime.

In the case of stainless steels, the first two categories give problems related to the high gas contents in the weld deposits. For practical purposes, the only types used are the basic and rutile coatings which both enable welding to be carried out positionally.

The exact compositions are rather variable, and every manufacturer adopts his own formula. However, coatings for stainless steels contain suitably balanced proportions of the following compounds, the figures quoted being only to give an indication of the constituents involved.

Basic coatings:

– Calcium or magnesium carbonates: 30 %.

– Calcium and, more rarely, magnesium fluorides: 20–30 %; this high proportion is due to the fact that the basic coverings have a high melting point.

– Alkaline carbonates, emissive oxides: 10 %.

– Clay, etc. (ensuring the solidity of the coating): some %.

– Alloying or additional metallic elements: 20 %.

– Binders such as alkaline silicates.

Basic electrodes can be used for all types of stainless steels. They have the advantage of permitting drying before use to be carried out at up to 300 °C, which allows the removal of all moisture from the coating. They give deposits having good mechanical properties, often with a rather convex shape, but the removal of the solidified slag is sometimes difficult.

Rutile coatings:

– Titanium dioxide (rutile): 40 %.

– Alkaline earth carbonates: 15–20 %.

– Calcium or magnesium fluorides: 10 % (the slags are more fluid than basic slags).

– Other compounds: as for basic coatings.

Rutile electrodes have the advantage of a higher arc stability, and a lower striking voltage than basic electrodes. They give beads with a good appearance, and slag removal is easy. They can give deposits of inferior quality to basic electrodes in some cases of positional welding. Further, they can withstand only moderate baking temperatures (200 °C), which can cause problems in welding martensitic stainless steels.

It should also be pointed out that, between these two main types, there are numerous intermediate variants, aimed at combining the respective advantages of basic and rutile coatings.

4.3 General comments

The covered electrode process has a number of distinct technological advantages. It can also have metallurgical advantages, particularly in the case of stainless steels. The high arc stability given by this process allows ready control of current and electrode movement with respect to the workpiece, so that the energy transmitted to the metal is low. In consequence, the degree of superheating of the weld metal is low, as is the heating of the base metal. These advantages will be considered further with reference to ferritic and austenitic steels.

4.4 Working conditions of stainless steel electrodes

Ordinary steel electrodes are used with alternating and, in particular, direct current. In the latter case, the positive pole of the power source is generally connected with the workpiece ('direct polarity'). This system has the advantage of giving good penetration as a result of the high liberation of heat at the anode, and is desirable because of the good thermal conductivity of ordinary steels. As direct polarity liberates little heat at the cathode, it is necessary to use rather high currents to promote fusion of the electrode. This is possible as the metallic core has a low electrical resistivity, ranging from about 1–2 $\mu\Omega$ mm (25–50 $\mu\Omega$ in.).

The situation regarding stainless steels is different, because the electrical resistivity of the wire is nearly 5–6 $\mu\Omega$ mm (125–150 $\mu\Omega$ in.) at room temperature for martensitic and ferritic steels, and 7–8 $\mu\Omega$ mm (175–200 $\mu\Omega$ in.) for austenitic steels. As a result, the welding current for a given core wire diameter must be about 20–30 % lower for stainless steels than for ordinary steels, in order to avoid excessive heating of the core wire. Moreover, to obtain good fusion characteristics, the positive pole of the power source is connected to the electrode ('reverse polarity'). Although less heat is generated at the workpiece, good penetration is obtained by virtue of the low thermal conductivity of stainless steels. It is also because of the electrical resistivity of the core that stainless steel electrodes are shorter than ordinary steel electrodes. It may be noted that rutile electrodes are also used with alternating current.

5 Submerged-arc welding

This process employs a bare wire acting as a consumable electrode, which melts under a flux deposited along the joint in the form of powder; in fact, the wire melts in a molten slag which is made electrically conducting by the high temperature.

From the point of view of the metallurgy of stainless steels, this process has the following characteristics:

– It causes a high loss of reactive elements. This loss must be compensated for in the same way as with covered electrodes.

– In most of its applications, the high specific energy involved causes significant superheating of the weld metal, and the formation of an extensive HAZ, while the presence of a slag, often as a thick layer, reduces the cooling rate of the weld. Moreover, the dilution (chapter 9, §2) is generally quite high, and can be inconvenient for certain categories of stainless steels.

The fluxes usually employed are of an acidic type (silicate predominating) or a basic type (lime predominating). This latter type is more common with stainless steels although, as in the case of covered electrodes, it must contain a substantial proportion of fluxing compounds. Acidic fluxes can also be used, but they may lead to a greater loss of oxidisable elements, and an increased hydrogen content of the weld metal.

Finally, mention may be made of the electroslag welding process used for the vertical welding of large pieces in only one pass. It has thermal characteristics similar to those of submerged-arc welding, with a substantial molten pool solidifying slowly, and is little used for stainless steels.

6 Gas-shielded processes

6.1 General comments

The covered electrode process is very flexible, but it has the disadvantages of being intermittent, and requiring a deslagging operation after each run, as is also the case with submerged-arc welding.

The gas-shielded processes do not have these drawbacks; they can be manual or mechanised, and can give welds of excellent quality, with freedom from inclusions and porosity.

Because of their chemical inertness, the gases employed are essentially argon and helium, or mixtures based on these gases. The former gas is used almost exclusively in Europe, where its basic cost is less than that for helium. The situation is different in the U.S.A., where there are natural sources of helium, and this gas is employed as far as possible.

Argon and helium ensure excellent protection of the molten metal against atmospheric gases, provided that turbulence is avoided in the arc region. In the downhand welding position, the protection for a given gas flow rate is more efficient with a denser gas, which favours argon. The pipes or tubes carrying the gas to the arc must be absolutely leakproof, of course, in order to avoid the entrapment of air or moisture from the cooling circuits of the welding torch.

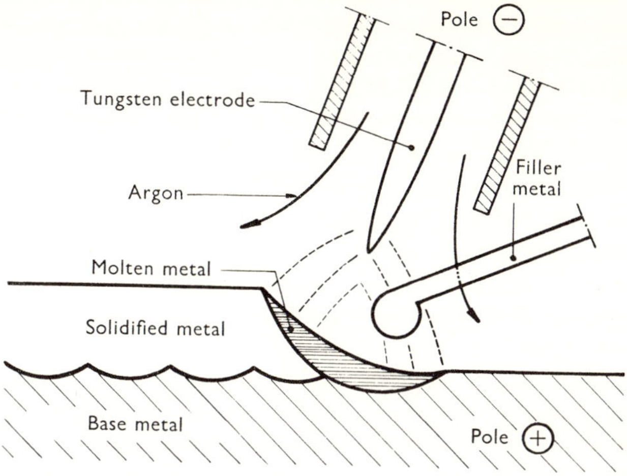

Fig. 3.7. Principle of the TIG process.

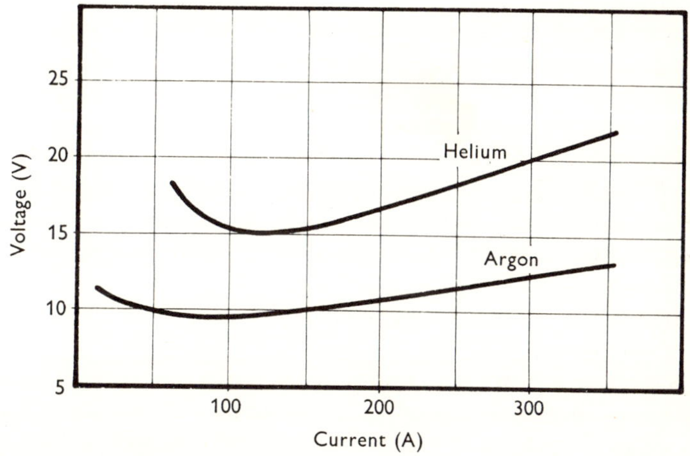

Fig. 3.8. Arc voltage for a distance of 2 mm (0·08 in.) between aluminium and a tungsten electrode under helium and under argon.

In conditions of good shielding, the chemical composition of the deposits is very close to that of the initial consumables, even when these contain highly oxidisable elements such as aluminium and titanium. It would, in fact, be impossible to compensate losses during fusion as is done with covered electrodes, because the filler metals employed in the gas-shielded processes are almost always bare wires.

Finally, it should be appreciated that these processes necessitate a more expensive material than the covered electrode process, and that with stainless steels, the high price of the shielding gas is also an important factor in the basic cost of making a weld.

6.2 Non-consumable electrode process

The principle of this process, often called 'TIG' (tungsten inert gas), is illustrated in fig. 3.7. The arc is struck between the workpiece and a tungsten electrode, which remains unmelted. The protecting gas forming the arc plasma is brought in through a nozzle which is concentric with the electrode.

The process as applied to stainless steels uses direct current, with the positive pole connected to the workpiece (direct polarity). This system allows most of the heat to be localised on the metal to be melted. In some applications, the use of a pulsed current enables certain characteristics of the process to be improved, such as arc stability and better penetration, and refinement of the weld metal microstructure [6].

The stability of the arc in the TIG process is ensured by ionisation of the shielding gas, and may be increased by the use of electrodes containing compounds which facilitate the emission of electrons, such as thorium oxide.

Argon and helium are not equivalent from the point of view of process interchangeability. Fig. 3.8 illustrates the arc voltages obtained with both these gases as a function of the current [7]; it is apparent that the arc voltage is higher with helium than with argon, and that the voltage minimum corresponds to a lower current in the case of argon than in that of helium. This difference is partly due to the higher ionisation potential of helium (24 V) than of argon (15 V). It should also be noted that both these gases can exist in metastable excited states, from which ionisation is easier; thus arcs can exist at voltages which are low in comparison to the ionisation potentials.

This difference of properties between argon and helium has the two following consequences:

– Argon is preferred for welding thin gauge materials, because the current at which the arc becomes unstable is lower than with helium. Further, variations in arc length cause higher energy variations with helium than with argon, which makes the latter preferable in manual welding.

– Helium is advantageous when a high welding speed (i.e. a high arc energy) is required, as in mechanised welding. The energy liberated at the anode increases with the total arc energy, and at equal current, helium furnishes a higher energy. Geometrical weld defects become apparent at similar current values for both gases, but the higher energy associated with helium permits higher welding speeds to be obtained.

It is possible to increase the arc voltage by adding hydrogen to argon. This is illustrated in fig. 3.9 [7], where it can be seen that 35 % hydrogen must be added to argon in order to obtain arc voltages similar to those obtained with pure helium.

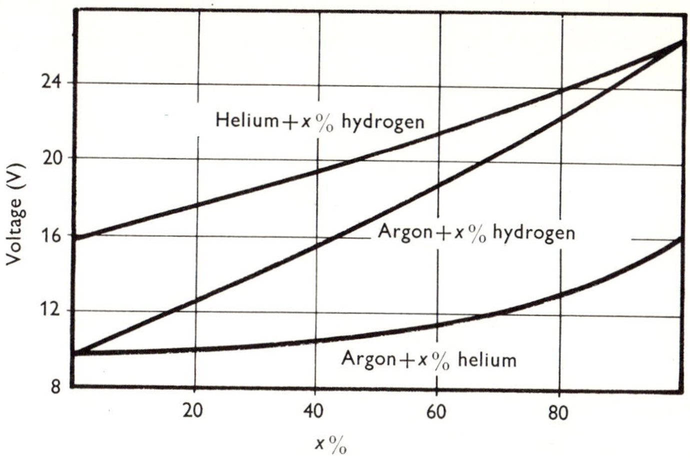

Fig. 3.9. Voltage for an arc of 100 A and a distance of 1·6 mm (0·065 in.) between stainless steel and a tungsten electrode, for different gas mixtures.

However, for the reasons given above (§1.2), it is advisable to limit the hydrogen proportion to 8 or 10 %, and to use this mixture only for the welding of austenitic steels.

Figure 3.9 also shows that argon–helium mixtures have only a limited interest from the point of view of increasing arc voltage unless the proportion of helium exceeds 60 %.

Finally, it should be noted that the use of a tungsten electrode, which is readily oxidised at high temperature, precludes the use of slightly oxidising gas mixtures, even though such mixtures may be advantageous in the case of the consumable electrode process (§ 6.5).

Depending on the thickness and shape of the assembly, the TIG process can be employed with or without additions of filler metal. It enables high quality welds to be made manually, but at a relatively high cost. Pure argon is employed, along with a power source having a very steep characteristic, so that the current is nearly independent of the voltage over the range of application. This gives good control of fusion. In mechanised form, the TIG process finds extensive industrial use for the fabrication of rolled and welded stainless steel tubes. An argon–hydrogen mixture is often used for austenitic steels.

6.3 Consumable electrode process

This process is also known as 'MIG' (metal inert gas). Its principle is illustrated in fig. 3.10. The electrode is a bare filler wire of diameter 0·6–2·0 mm (0·024–0·080 in.), fed at constant speed through a copper tube to obtain electrical contact. The arc is struck between the workpiece and the wire, which melts at an average rate equal to the feed speed as

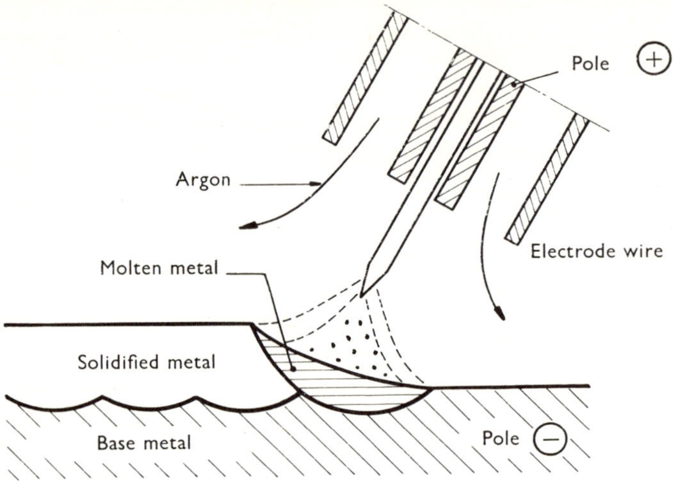

Fig. 3.10. Principle of the MIG process.

controlled by suitable selection of the power source characteristic. The inert gas is brought in through a concentric nozzle around the electrode.

As the mode of transfer of the liquid metal with direct polarity is quite different from that with reverse polarity, only direct current is used. When argon is employed, metal transfer at above a certain current level takes place regularly in the form of very fine drops. These drops are violently projected towards the molten pool, provided that the consumable electrode is connected to the positive pole of the power source (reverse polarity), and this transfer mode is known as 'spray transfer'.

The general mechanism of metal transfer has been briefly reviewed above (§ 3.2). However, it is necessary here to consider the phenomenon of polar spots, which causes certain irregularities in transfer across the arc. It is known that the current density is not uniform across the entire surface of an electrode [4], and that at any moment, the current flows through only a very small region termed the 'anode spot' or 'cathode spot', as the case may be. This localisation of current flow corresponds to a minimum expenditure of energy at every instant, and according to the geometrical development of the system, polar (cathodic or anodic) spots move very quickly over the surface of the electrodes to establish this condition of minimum energy.

Through the polar spots, electrons and eventually ions are emitted, resulting in a repulsive force which is particularly marked at the cathode. This effect is responsible for the globular mode of transfer found in an argon atmosphere when the negative pole of the power source is connected to the electrode. The repulsive force due to the cathode spot opposes the detachment of the molten drop from the wire; the drop separates only when its volume becomes large, and it then follows a complicated trajectory to the molten pool. Moreover, only moderate heat is liberated at the cathode, and the viscosity of the metal facilitates this mode of transfer.

Globular transfer occurs with helium whatever the electrode polarity. It is thought that the repulsive effect of the anode spot becomes significant in this case, because of the high thermal conductivity of helium, which cools the anode and increases the metal viscosity, and as a result of the low gas density which reduces the streaming effect of the plasma. The good transfer characteristics obtained with pure argon are regained when more than 20 % argon is added to helium, though the molten metal drops remain bigger than with pure argon. This mixture also allows advantage to be taken of a slightly higher arc voltage. Unlike the situation with TIG welding, inert gas–hydrogen mixtures do not give good results from the point of view of arc stability.

With positive electrode and argon shielding, it is found that as the current is reduced, there exists a threshold below which the metal transfer becomes globular. The use of argon containing 1–5 % oxygen considerably lowers this limit and, as a whole, improves the arc stability with stainless steels as well as ordinary steels. Oxygen seems to reduce the surface tension and viscosity of the molten pool, thus facilitating detachment of the drops by the pinch effect.

The addition of oxygen to argon also has a favourable effect on the weld profile, which can be explained on the following basis. When an arc is struck under argon, the cathode (workpiece) is bombarded by relatively heavy ions; this bombardment has a cleaning effect which helps the wetting by the molten metal. Although oxygen opposes this effect, it lowers the viscosity of the molten metal, and hence favours the formation of a smooth bead with limited dimensions.

Further, the addition of oxygen results in more uniform penetration. With pure argon, which is a poor conductor of heat, the axis of the arc plasma is much hotter than the peripheral region, and the penetration is localised at the centre of the weld. Since oxygen increases the thermal conductivity of the plasma, it reduces the differences in penetration, particularly with stainless steels which also have a low thermal conductivity.

It is possible to obtain similar results by replacing oxygen with carbon dioxide which also has an oxidising effect; oxygen is, however, most commonly used. It is advisable not to add too much oxygen, to avoid oxidation of the molten metal or the formation of porosity.

For these reasons, the MIG process applied to stainless steels generally employs argon with 1 % oxygen. It is possible that argon–helium mixtures could become more widely used in situations where their high basic cost does not constitute a disadvantage.

For nickel-base alloys, pure argon is used to avoid porosity as well as a certain type of hot shortness.

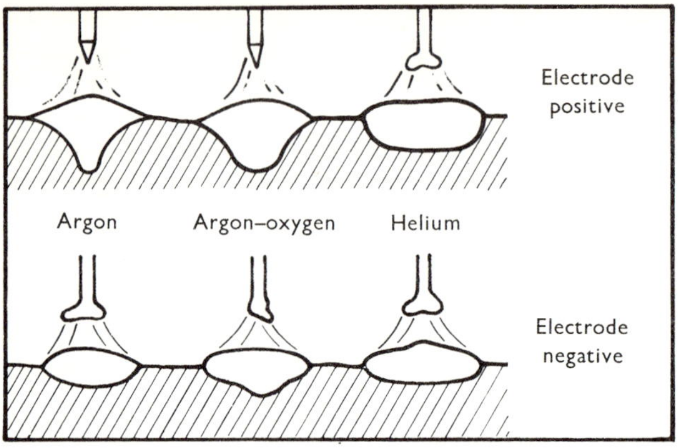

Fig. 3.11. Possible configurations in MIG welding.

The use of these gases is costly, and this has led for the gas-shielded metal-arc welding of ordinary and low alloy steels to the use of mixtures of argon with more than 15 % carbon dioxide, mixtures of argon–oxygen–carbon dioxide, or almost pure carbon dioxide atmospheres.

The use of carbon dioxide for the welding of stainless steels is not advisable for the following reasons:

– At the very high temperatures found in the arc column, the gas is partially dissociated into oxygen and carbon monoxide, creating an oxidising atmosphere that acts mainly on the metal during its transfer. This results in severe oxidation in ordinary steels, with a risk of porosity in the deposited metal due to reaction between carbon present in the steel and the iron oxide; this problem is reduced by the use of filler wires containing silicon and manganese, which form oxides which are less readily reduced by carbon.

– At lower temperatures close to the molten pool, the gas and its dissociation products are in equilibrium with the carbon of the steel; the result is that the molten metal under carbon dioxide always contains about 0·1 % carbon, nearly independent of the initial carbon contents of the filler wire or workpieces. This is advantageous for ordinary steels, but often undesirable in the case of stainless steels.

It has been pointed out above that the direct polarity used with pure argon leads to a globular transfer. As a result, there is no 'impact energy' as associated with a fine spray transfer, and the penetration is relatively low. This system can be employed to overlay an ordinary steel with stainless steel, when low dilution is required as considered in chapter 9. Practically, the different cases considered are schematised in fig. 3.11, summarising the transfer modes and penetration characteristics [7].

The current also has a marked effect upon the transfer mode. It has already been pointed out that there exists a limit below which the transfer

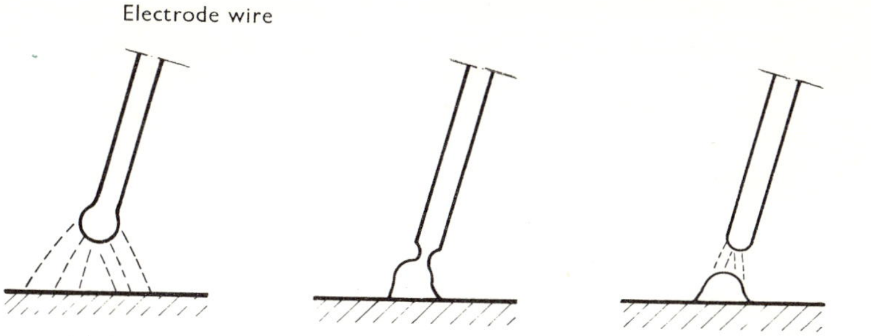

Fig. 3.12. Principle of the periodic short-circuiting process.

becomes globular, probably because of an increase in the molten metal viscosity. As the current decreases, larger and larger drops form, which may even short-circuit the arc. When technological conditions regarding the wire feed rate and the static and dynamic characteristics of the power source are fulfilled, the transfer mode, by periodic short-circuits, becomes stable, and gives rise to a variant of the process known by its original terminology as the 'short arc process', 'fine wire process', 'dip-transfer', and by certain commercial names.

The sequence is illustrated in fig. 3.12. At the beginning of the cycle (fig. 3.12(*a*)) an arc exists, but the wire feed rate is too high with respect to the current. The arc length decreases, while a molten metal drop forms at the wire tip. A short-circuit therefore occurs before the drop becomes detached (fig. 3.12(*b*)). At this moment, the current increases sufficiently to ensure metallurgical bonding between the plate and the drop, and to break the contact by a pinch effect on the molten metal, although without causing an explosion of the molten drop (fig. 13.12(*c*)). An arc is re-established at a current which is still very high, leading to rapid melting of the wire. The arc length quickly increases until the situation in fig. 13.12(*a*) is attained again, and the cycle recommences.

The spray transfer process as applied to stainless steels, uses wires having a diameter of 1–2 mm (0·04–0·08 in.) with relatively high currents, for instance of the order of 300 A for an austenitic steel wire with a diameter of 1·2 mm (0·048 in.). It gives a high welding speed, but causes the formation of a large molten pool with high dilution (chapter 9, § 2). The heat input at the workpiece is considerable, and thus the cooling rate of the weld is relatively slow. The process can be employed in manual or mechanised form, but does not allow welding in other than the downhand position.

The advantage of the short-circuiting process is that it gives a high deposition rate at a low average current, and so with a low heat input. Moreover, the size of the molten pool is reduced, and the pool cools

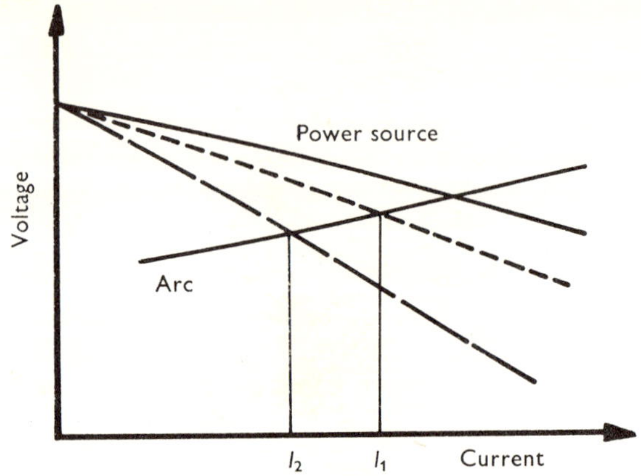

Fig. 3.13. Arc and power source characteristics. ----- arc voltage with ordinary steel wire; ——— arc voltage with stainless steel wire.

rapidly, permitting welding in the vertical or overhead positions. It is employed mainly in its manual form, with wires having a diameter from 0·6–1·2 mm (0·024–0·048 in.).

The two variants of the MIG process utilise welding power sources with very slightly drooping volt–ampere characteristics, allowing automatic compensation for fluctuations in the consumable wire feed rate. With this system, an increase in the wire feed speed, which tends to decrease the arc length, causes a slight voltage reduction and a high increase in current; this raises the burn off rate of the wire until a new equilibrium is attained. This has the further result that, once the power source characteristic has been selected, the arc length and current, and so the transfer mode, depend only upon the wire feed rate.

The working stability is reduced when the slope of the power source volt–ampere characteristic becomes excessively drooping. Now, the voltage across the arc is equal to the open-circuit voltage minus the voltage drop in the stickout of the wire. This voltage drop is not negligible since the wire diameter is small and the currents are high, and it is much more important with stainless steels than with ordinary steels, as shown in fig. 3.13. The result is that a MIG power source which is satisfactory for ordinary steels may be unsuitable for stainless steels. Though such technological considerations go beyond the general outline of this book, it is important to emphasise the fact that the MIG welding of stainless steels or nickel-base alloys is best carried out with a power source designed for this type of alloy.

6.4 Metallurgical variables affecting the use of the gas-shielded processes

Fabrication operations based on rolling or forging of stainless and heat-resisting steels are more difficult than for ordinary steels. In particular, the hot-working of austenitic alloys on a reasonably economic basis is possible only when these metals are completely deoxidised. Thus, during their production, small quantities are added of elements having a great affinity for oxygen, such as alkaline earth metals, rare earth metals, aluminium, titanium and zirconium. Some of these elements have low ionisation potentials, and small amounts of such elements can remain in the metal either in solid solution, or in the form of non-metallic compounds. Their presence can cause irregularities in the functioning of gas-shielded arc processes under the following conditions:

– When these elements are volatile at the arc temperature, or form non-metallic compounds which boil at a temperature below that of the arc, their low ionisation potential reduces the arc voltage and consequently the energy appearing in the form of heat at the electrodes.

– When these elements are found in the form of non-metallic substances which are liquid at arc temperatures (and so conducting), they can form stable polar spots.

Calcium and the rare earth metals, in particular, fulfill these conditions.

These perturbations have different consequences, depending upon the process being considered. However, in all cases they are significant only when the process is mechanised, since with manual welding the welder compensates for such irregularities by movement of the welding torch.

TIG welding. In the current practice of this process, the positive pole is connected to the workpiece. When a low ionisation potential element is in solution in the metal, its volatilisation causes a reduction in the voltage drop at the anode, and this necessitates an increase in the welding current to obtain good penetration.

The effect is similar when the element is present in the form of non-metallic material which vaporises at the arc temperature. However, in this case the non-metallic substance may also gather in the form of a liquid drop at the surface of the molten pool, so that it fixes the position of the anode spot. When this drop is unable to follow the arc, for instance because it becomes attached to the edge of the bead, the energy dissipated at the anode suddenly rises again to its normal value; a new drop of non-metallic material begins to form and the cycle restarts. This results in an instability in the welding operation, and welding defects and excessive penetration follow one another.

Similar perturbations can occur when elements having a low ionisation potential appear at the cathode although the polarity is reversed. Irregularities in the weld can be even more pronounced, as less heat is liberated at the cathode.

The penetration of the weld bead depends upon the base metal composition. As an example, the addition of Si, Mn and Ti to the nickel alloy Inconel 600 enables both the penetration and quantity of weld metal to be increased. The simultaneous addition of certain elements can lead to complex interactions.

These additions generally have the effect of changing the electrical characteristics of the arc. Thus, during the TIG welding of nickel with direct polarity, the arc potential corresponding to a given current (200 A) is increased by additions of Al, Ca, V, Cr, Sn, Mn, Cu, Fe, Si and Sb. Other characteristics of the arc may be modified at the same time [8].

Various other factors, such as the purity of the shielding gas, are liable to influence the penetration. In the example given above for Inconel 600, if impurities such as N_2, O_2 in the gas increase by 0·07 % to 0·25 %, the penetration is decreased by about 50 %. This effect has been related to the increased emission from compounds formed from the impurities (especially oxides); the secondary electrons emitted in this way tend to retard the primary electrons emitted by the cathode, hence causing a decrease in penetration [9]. It has been possible to establish a correlation between the size of the anode spot and the quantity of easily ionisable impurities.

In the same context, the volatisation of alkaline or alkaline earth elements present in the metal, mainly in the form of non-metallic inclusions, can lead to a reduction in the arc voltage, to an increase in the size of the anode spot, and consequently to a decrease in the penetration. This effect can be reproduced by sprinkling such compounds on plates which, in their absence, give welds with satisfactory penetration.

MIG welding. When elements or compounds which are volatile at the arc temperature are present in the filler metal or workpiece, their easy ionisation has an effect comparable to that of a decrease in current, i.e. the metal transfer tends towards a possibly unstable globular mode. When the low ionisation potential component is present in the wire, the formation of a stable anodic spot also contributes to this phenomenon by creating a repulsive force, and spray transfer can become impossible.

This transfer mode is detrimental in itself to the weld penetration. Moreover, a marked decrease in the energy liberated at the cathode (workpiece) restricts the penetration to a narrow region, and this results in poor wetting of the base metal by the liquid metal as shown in fig. 3.14. This effect of poor wetting is even more pronounced when the low-ionisation potential element is present in the plate being welded.

The use of a highly deoxidised filler metal can also contribute to difficulty in wetting, because the favourable influence that oxygen would have upon the fluidity of the molten metal is lost.

As indicated above, low ionisation potential elements are especially detrimental in TIG welding when they are in the form of non-metallic material which is fusible and volatile at the arc temperature, as it is in this case where irregularities in penetration are observed. The presence of volatile elements in solution is less troublesome, because the penetration remains uniform, although it is necessary to increase the current. The

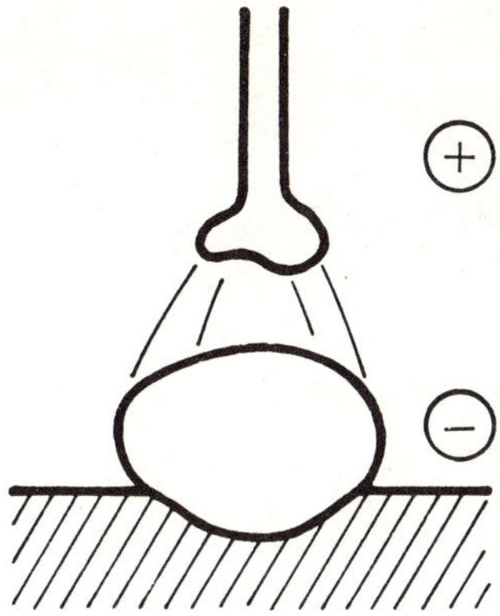

Fig. 3.14.

situation is different with MIG welding, because an unfavourable effect is manifest whatever the form in which the low ionisation potential element is found. Thus the presence of traces of cerium (0·009–0·018 %) in stainless steel electrodes causes irregularities in penetration as mentioned above, accompanied by a lack of fluidity and poor wetting. Cerium oxide inclusions, which are thought to be the cause of these difficulties, are in fact distributed unevenly through the bulk of the metal.

Silicon, on the other hand, has a favourable effect, as pointed out earlier, and this can be related to the increase in weld pool fluidity which this element causes [10].

6.5 Surface defects

Apart from the influence which they can have on the arc behaviour, non-metallic materials in stainless steels can affect the weld bead appearance. During the formation of the molten pool, these substances decant rapidly and gather at the bead surface as a result of convection in the liquid metal, and when their volume is sufficient, they adhere to the edge of the bead. If they are liquid at arc temperature, they form a glassy lens (fig. 3.15); if they are refractory, they form a 'puckered' surface which can take different forms (fig. 3.16).

Further, in titanium-stabilised austenitic steels, the nitrides or carbo-nitrides of titanium which are found in the steel can also form characteristic figures resembling spear heads on the weld bead.

The presence of these lenses or 'puckerings' has no influence upon the service properties of welds, as this phenomenon is only superficial. However, it can be inconvenient in causing surface irregularities or defects, in view of the qualities of appearance which may be required for finished products in stainless steel. It should be appreciated that these defects are not peculiar to stainless steels, but they escape notice on ordinary steels.

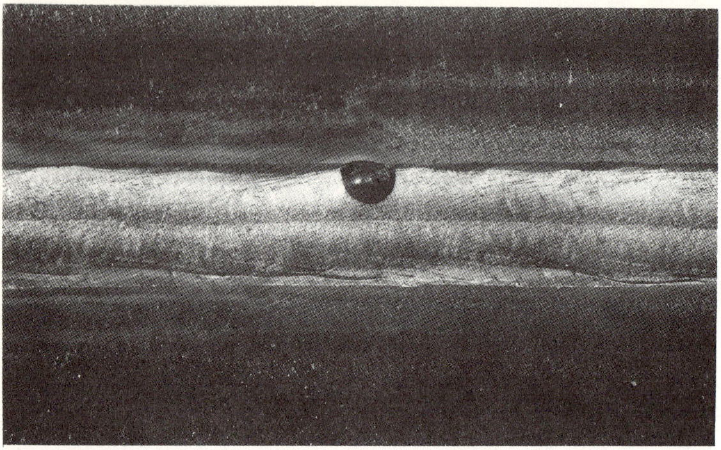

Fig. 3.15. Glassy lens on a TIG weld in austenitic steel (× 2·5).

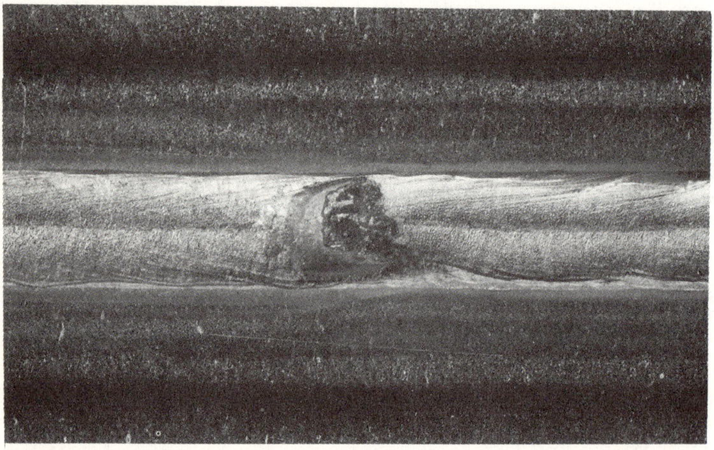

Fig. 3.16. Puckered surface on a TIG weld in austenitic steel (× 2·5).

Furthermore, these defects are important only in mechanised welding processes, as manual procedures cause surface irregularities of a comparable importance on the beads.

Remarks. The presence of low ionisation potential elements or non-metallic substances in stainless steels is directly related to the raw materials employed for the metal production, or to the production process itself. Their removal is either impossible or incompatible with reasonably economical manufacturing conditions. As is often the case, a solution must be found in a compromise between the service requirements, and the necessities of metal fabrication.

6.6 Other processes

Consumable electrode process with pulse controlled transfer [11], [12] (*pulse MIG welding*). It has been shown that numerous factors can influence the transfer mode in the gas-shielded metal-arc process. The number of these factors, and the relatively critical influence which they can have with stainless steels, necessitate investigation into procedures to control the transfer mode by electrical means. The following method, which is not fully developed, is of interest. A low direct current is used giving globular transfer, with the regular superimposition of sufficiently high current impulses to produce spray transfer during short periods. By automatically controlling the size of the molten metal drops, this allows the use of a low average current, thus avoiding the irregularities in penetration and bad shape caused by an unstable transfer mode.

'*Flux cored*' *process*. In this gas-shielded process, the consumable wire is a rolled strip in the form of a tube containing fluxing products and often powdered metallic elements. It is comparable to the covered electrode process, particularly in view of the possibility of obtaining deposits of different compositions from the same strip.

Plasma process. This process is related to the TIG process, but the heat is provided by a much more ionised, and thus much hotter plasma jet than the conventional arc. This is obtained by constricting the electric arc, using a gaseous jet of a suitable shape, and particular arc conditions. The workpiece can be connected to a pole of the power source ('transferred plasma') or be electrically neutral ('non-transferred plasma').

With stainless steels, the plasma is formed by pure argon or argon–hydrogen mixtures. The advantage of this process applied to stainless steels lies in the extreme concentration of the heat liberated, which permits its use for the welding of massive pieces as well as thin pieces.

Atomic hydrogen process. The TIG process is derived from this method. Heat is liberated at the workpieces by the recombination of hydrogen into molecular form, following dissociation in an arc maintained between two refractory electrodes. Despite the nature of the atmosphere surrounding the weld, this process is only moderately hydrogenising as there is no polar (cathodic or anodic) spot on the molten pool. It gives welds of a very good appearance.

Electron beam process. In this process, the fusion of the metal is obtained by an electron beam, the electrons being produced by an electron gun and focused by electromagnetic lenses. The welding operation takes place in vacuum or in a rarefied inert atmosphere. The process has the advantage of giving deep penetration, a very limited HAZ, excellent protection of the metal and an efficient degassing of the molten metal. It is applied to the welding of difficult workpieces, for which distortion after welding must be minimised, or of metals sensitive to cracking in the

HAZ[13]. It has been possible to carry out welding at pressures reaching one atmosphere, but the focusing of the electron beam as well as the penetration are less satisfactory, and the bead has a tendency to spread.

Laser process. Here the source of energy is a coherent light beam emitted by a laser. This process, which has a much lower energy input than the preceding one, is applied to the welding of very thin wires or plates[13].

7 Resistance welding processes

In these processes, the molten metal arises only from a partial fusion of the pieces to be assembled, and the operational cycle involves at least one period during which the workpieces are strongly pressed against one another. Only processes using electrical heating will be considered.

7.1 Resistance welding of sheets (spot welding)

The principle of this process is shown in fig. 3.17. The plates to be welded are overlapped and pressed together. This period, known as the 'forging time', allows break up of the surface layers of the metal, giving reduced contact resistance. Current then flows in the workpieces, while the pressure exerted by the electrodes is maintained at a level which may be different from that during forging. This is the 'welding time', during which a common nugget of fused metal is formed between the two plates being welded.

The localisation of this molten nugget can be explained on the following basis. At the beginning of the current flow, the temperature distribution has the form shown by the dashed curve in fig. 3.17[14], [15], because the interfaces have the highest electrical resistance and thus are the centre of the maximum heat liberation. The surfaces in contact with the electrodes are heated least, since the electrodes are good conductors of heat, and are often water-cooled. The preferential heating of the central region is self-accelerating because the resistivity of the metal increases with its temperature. The result on completion of welding is a temperature distribution having the shape of the solid curve in fig. 3.17, thus causing the formation of the molten nugget constituting the final spot weld.

During solidification and cooling, the molten nugget is completely surrounded by solid metal, which creates a risk of piping and cracking due to contraction. To obtain a sound spot weld, the pressure exerted by the electrodes is maintained or increased during cooling (the 'holding time'). This pressure gives rapid extraction of the welding heat, and forging of the weld.

The application of this process to stainless steels is characterised by the fact that all these materials have a high electrical resistivity, a low thermal conductivity, and a high contact resistance. The result is that the self-accelerating character of the fusion process is much more pronounced than in the case of ordinary steels. In consequence:

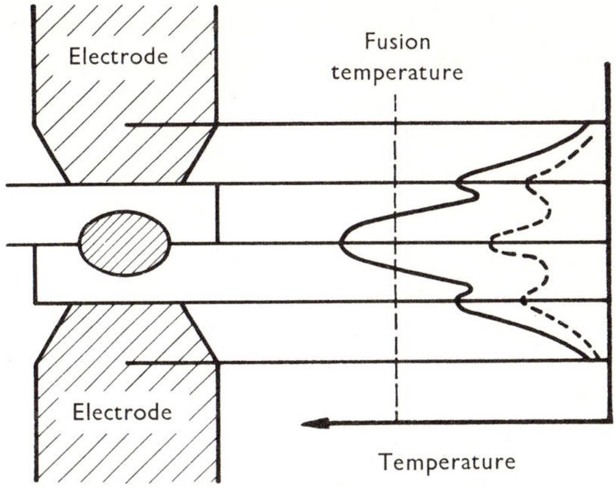

Fig. 3.17. Principle of resistance welding for sheet material.

- The electrode pressure should be increased to reduce the contact resistance and facilitate removal of heat through the electrodes.
- The current and welding times should be reduced.
- The welding currents and times must be controlled more accurately than with ordinary or low alloy steels.

These comments are equally valid for seam welding, which consists of a succession of spots being partially overlapped to form a continuous joint.

7.2 Resistance butt welding of bars and sections

The principle is similar to that for sheet welding. A forging phase and a welding phase can be distinguished, during which the heat is produced mainly in the vicinity of the abutting surfaces of the pieces to be assembled. In resistance butt welding, the welding phase includes an upsetting of the workpieces. This gives expulsion from the centre of the joint of some molten metal, and in particular of oxides which are formed at the abutting surfaces during the initial heating period. The operation finishes by holding the components together under pressure which has the same effects as in spot welding, apart from cooling. In the case of ferritic steels, it can be advantageous to make the weld by forging rather than by fusion, as considered in chapter 5, §7.

7.3 Flash butt welding of bars and sections

In this process, the pieces to be assembled are placed together under light pressure, and an electrical potential is applied between them. Current flow takes place through asperities in contact, giving very high current densities, which cause instant fusion and explosion of liquid droplets. The points through which the current flows rapidly move over the abutting surfaces, and the pieces are slowly brought together in such a way that this type of functioning remains stable. The contact points become more and more numerous, and the local original current densities diminish. In this way, a very superficial layer of metal is melted while the regions adjoining the surface are heated by conduction. This is the flashing period.

The pieces are then strongly pressed together, possibly at a reduced current. This results in considerable upsetting of the heated material, which expels molten metal and oxides from the useful section of the joint, and also affects a part of the remaining solid metal.

The process is characterised by a very localised heating, a heavy forging, and an absence of metal which has been melted in the centre of the joint.

7.4 Induction welding

This process is mainly employed in the case of stainless steels for the fabrication of longitudinally rolled and welded tubes. The edges to be assembled are heated by induction with an average or high frequency, using an inductor having a special shape, and are then pressed strongly against each other to make the weld. From a metallurgical point of view, this process can be considered to have effects intermediate between those of resistance welding and those of flash welding.

References

[1] D. J. Carney. *Gases in metals.* A.S.M. 1953.
[2] A. Leroy, M. Evrard and G. d'Herbemont. *La pratique du soudage oxyacétylénique et des techniques connexes.* Publications de la Soudure autogène, Paris, 1963.
[3] A. Gaubert. *Soudage Techniques connexes*, **13** (1959) 85.
[4] A. Almar Naess. Document 115/1WS 129–64. *Soudage Techniques connexes*, **10** (1965) 20.
[5] Norme française A 81–309, 1962.
[6] E. P. Vilkas. *Weld. J.* **49** (April 1970) 255.
[7] *Welding Handbook*, AWS, Section 2–27.
[8] Th. F. Chase and W. F. Savage. *Weld J. Res. Sup.* **50** (Nov. 1971) 467s.
[9] W. F. Savage, C. D. Lundin and G. Goodwin. *Weld. J. Res. Sup.* **47** (Jul. 1968) 313 s.
[10] C. E. Cassel. *Weld. J. Res. Sup.* **48** (April 1969) 167.
[11] P. Boughton and J. A. Lucey. *Br. Weld. J.* **12** (1965) 159.
[12] J. C. Needham. *Br. Weld. J.* **12** (1965) 191.
[13] R. P. Sullivan. *Metals Eng. Quart.* (Nov. 1967) 16.
[14] *Welding Handbook*, AWS Section 2–27.
[15] J. Nègre. *Le soudage électrique par résistance.* Publications de la Soudure autogène, 4th ed, Paris, 1965.

4

WELDING OF STAINLESS AND HEAT-RESISTING MARTENSITIC STEELS

1 General comments: composition and structure

It was pointed out in chapter 1 that stainless and heat-resisting martensitic steels are related to iron–chromium alloys, with such a composition that, while cooling from the solidus, they pass through a temperature range in which they are completely transformed from delta ferrite into austenite. In addition to the main elements, iron and chromium, martensitic stainless steels contain minor alloying elements such as carbon, silicon, manganese, nitrogen, nickel and possibly other elements, as considered below. In the most common commercial martensitic alloys, silicon, manganese, nitrogen and nickel are kept at relatively low and constant levels. The carbon content, however, may be altered over a wide range, since this element permits considerable variation to be obtained in the mechanical characteristics of martensitic steels.

To a first approximation, the physico-chemical behaviour of martensitic stainless steels is similar to that of hardenable carbon steels. Like the latter, they solidify as delta ferrite which transforms into austenite during cooling. When the cooling is sufficiently slow, this austenite is decomposed into ferrite and carbides owing to a eutectoid reaction. If the cooling is fast enough, the austenite forms martensite.

The presence of chromium results in a shift in the carbon content corresponding to the eutectoid point towards lower values. It also reduces the carbon solubility, and causes the formation of three types of carbides containing iron and chromium. Consequently, this gives a rather complicated metallurgy, in which the alloy properties vary rapidly as a function of carbon and chromium content, as well as with heat treatment.

Figure 4.1[1] shows vertical sections of the equilibrium diagram of industrial Fe–Cr–C alloys containing 0·05 %, 0·10 % and 0·20 % carbon. These sections, which clearly show the similarity to carbon steels, also permit the following observations to be made: above approximately 7 % chromium, an increase in the chromium content decreases the extent of the temperature range over which austenite is stable. In practice, a purely austenitic structure (or austenite + carbides) exists only below 12 % chromium with 0·05 % carbon, below 13·5 % chromium with 0·10 %

carbon, and below 15·5 % chromium with 0·20 % carbon. These limits define the usual compositions of martensitic stainless steels.

The effect of carbon is opposite to that of chromium, in that, at constant chromium content, an increase in the carbon content widens the temperature range over which austenite is stable. At constant temperature, a rise of the carbon content widens the austenite range to higher chromium contents. This is apparent in fig. 4.2, showing the section of the equilibrium diagram for a commercial 13 % chromium steel. This is the most widely used type of martensitic stainless steel, and it becomes fully austenitic only when its carbon content exceeds about 0·08 %.

In practice, all these steels are self-hardening, which means that the austenite is not transformed into alpha ferrite and carbides during air cooling from the austenite range. The austenite remains stable under these conditions down to a temperature known as the 'M_s' point, below which a diffusionless transformation to martensite occurs; this martensitic transformation is completed at a temperature known as 'M_f'. For a steel containing 0·12 % carbon and 12 % chromium, the M_s and M_f temperatures are respectively 300–350 °C and 150–180 °C. These temperatures decrease when the chromium content is increased, and when other alloying elements are present in the steel.

Martensitic stainless steels are generally employed in the quenched and tempered, or the annealed condition. In the first case, quenching causes the

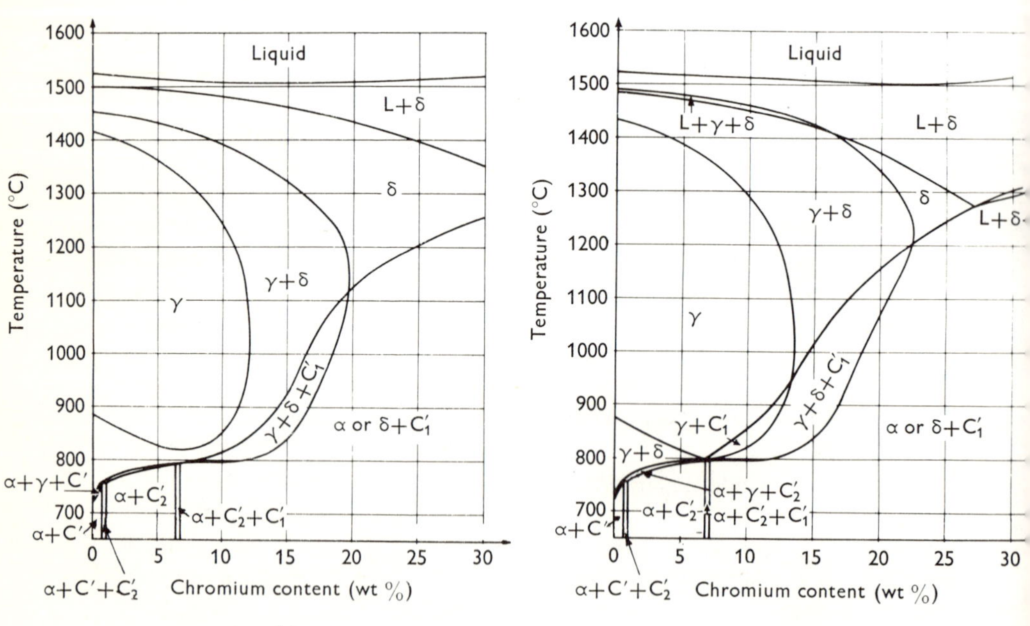

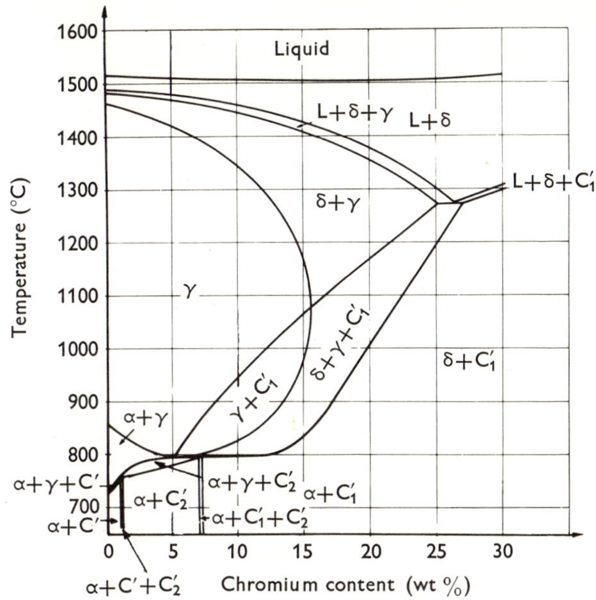

(c)

Fig. 4.1. Vertical sections of the equilibrium diagram for industrial Fe–Cr–C alloys: (a) 0·05 % C, (b) 0·10 % C, (c) 0·20 %C.
(Ni ⩽ 0·13 %/N ⩽ 0·013 %/Si ~ 0·4 %/Mn ~ 0·5 %)

C′ : carbide $(Fe, Cr)_3 C$ — L : liquid
C′₁ : carbide $(Cr, Fe)_{23} C_6$ — δ : delta ferrite
C′₂ : carbide $(Cr, Fe)_7C_3$ — γ : austenite
α : alpha ferrite

formation of a hard martensitic structure with little ductility, which is decomposed into ferrite and finely dispersed carbides by a tempering operation carried out at below the austenite range. The choice of the quenching and particularly the tempering temperature allows various combinations of mechanical strength and ductility to be obtained, as shown in fig. 4.3. Annealing is performed at a temperature in the austenite range, and is followed by a sufficiently slow cooling for austenite to be decomposed into ferrite and carbides. Annealing gives the maximum softening for wrought steels.

Martensitic chromium steels are employed because of the combination of their mechanical properties with their chemical inertness. This latter property permits them to resist corrosion for use in normal atmospheric conditions, in soft water (hydraulic turbines), in contact with food products, and in the chemical industry, especially the petroleum industry.

Their mechanical properties at high temperature find application for

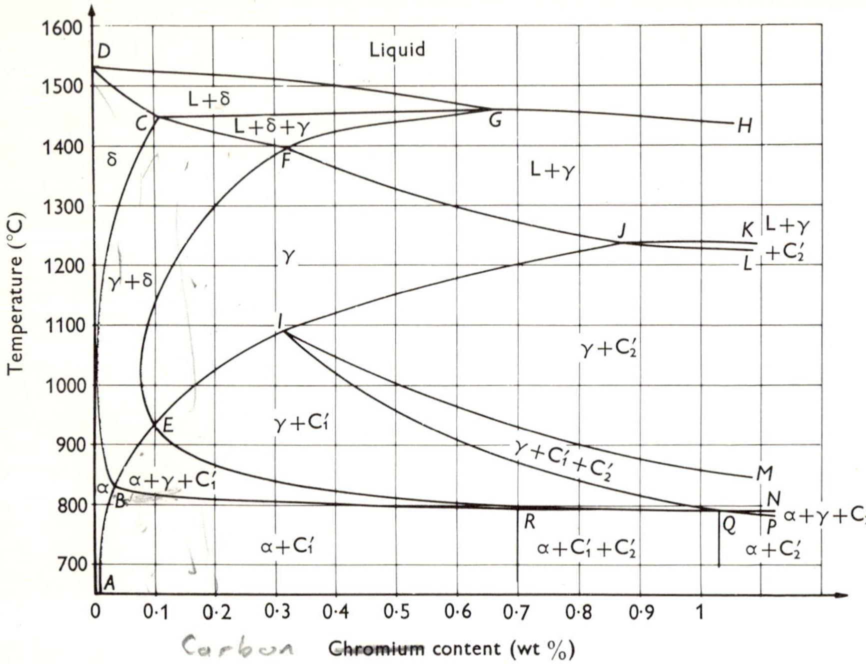

Fig. 4.2. Vertical section at 13 % chromium of the diagram for Fe–Cr–C alloys in fig. 4.1.

heat exchangers, steam turbines, valves etc. For this type of service, the basic martensitic chromium steels may be modified by addition of small quantities of molybdenum, tungsten, vanadium, niobium etc., which are mainly intended to improve the creep resistance. These elements have a physico-chemical effect comparable with that of chromium, and as the carbon content is limited for reasons of weldability (see below), the chromium content is reduced so that these steels remain fully martensitic with no delta ferrite, their good mechanical properties being related to the martensitic structure.

For the same reason, the chromium content cannot be raised to improve the corrosion resistance without impairing the mechanical properties, unless an element other than carbon is added promoting austenite stability. This has led to the development of a range of steels containing 16–20 % chromium and 0·1 % carbon which are rendered martensitic by the addition of 2–4 % nickel. The metallurgy of these steels is similar to that of conventional martensitic alloys, but their corrosion resistance, particularly in sea water, is improved by their higher chromium and lower carbon contents.

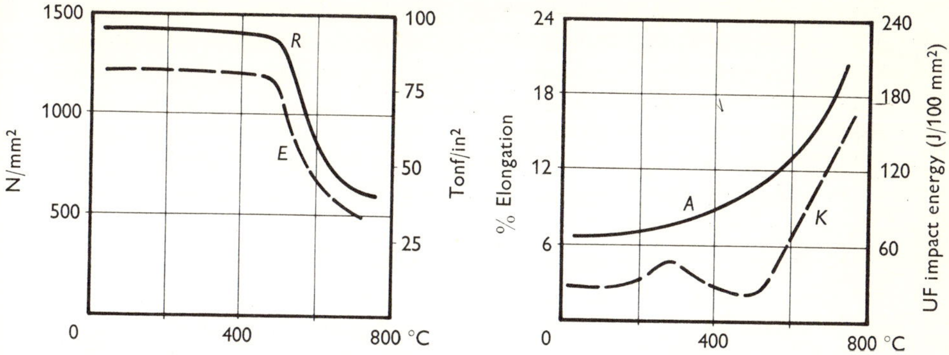

Fig. 4.3. Effect of the tempering temperature after quenching on the mechanical characteristics of a 13 % Cr/0·1 % C steel: *R*, tensile strength; *E*, elastic limit; *A*, elongation; *K*, toughness – UF impact energy.†

2 Microstructural changes on welding

Considering the case when the weld metal has the same composition as the base metal, the weld structure after cooling results from the following transformations.

2.1 Solidification and transformation to austenite

Solidification takes place as delta ferrite, which subsequently transforms to austenite. It has been shown above that complete transformation takes place only if the chromium and carbon contents are suitably adjusted, taking into account minor or other alloying elements that may be present. Silicon and molybdenum in particular promote the formation of stable delta ferrite. This is also the case for aluminium, which is considered below in connection with ferritic and semi-ferritic steels (chapter 5). By forming very stable carbides, niobium and vanadium have an effect comparable with that of a decrease in carbon content, and their presence requires a reduction in chromium content so that the fully austenitic range can be traversed on cooling.

However, the transformation to austenite can be incomplete even if the alloy has an appropriate composition from the point of view of the equilibrium diagram, for the two following reasons: Firstly segregation

† The Mesnager U-notch test specimen has a variable notch depth, and in France, the test results obtained are increased by a suitable factor so that they are applicable to a cross-sectional area of 100 mm² at the notch root. Because this factor is variable, the UF impact energies cited in the book have not been converted to Imperial units since impact energies quoted in the UK do not usually recognise variations in notch depth. SI units have, however, been adopted. The Charpy V-notch specimen is of fixed dimensions, and impact energies derived from Charpy tests have been converted to Imperial units on the basis of a cross-sectional area of 80 mm², i.e. the data in the original French have been multiplied by a factor of 0·8. The original French V-notch energies are in da J/cm²; these have been converted to SI units of J/100 mm² (1J = 0·737 ft lb.).

during solidification can cause the formation of ferrite, which remains stable during cooling. For instance, fig. 4.2 shows that the metal solidifying first is impoverished in carbon as compared to the liquid metal. This tends to stabilise the ferrite corresponding to the dendrite axes. Secondly, though very fast, the transformation of ferrite into austenite during cooling is more complete as the cooling rate is lower. As the cooling rate of a weld at elevated temperatures is high, some non-equilibrium delta ferrite can remain.

On the other hand, during solidification, the remaining liquid may become enriched in ferrite-stabilising elements, and this can render the ferrite corresponding to the inter-dendritic regions stable during cooling. Therefore, it is not impossible to observe the presence of small quantities of ferrite simultaneously within and between the dendrites [2].

Generally, it is not desirable for the weld to contain too much delta ferrite. It is impossible to control the mechanical properties of this phase by heat treatment; moreover, as indicated in chapter 5 this phase is very notch-sensitive, and if present to any marked extent it can impair the toughness of the weld after heat treatment.

2.2 Martensitic transformation

Martensitic chromium steels are self-hardening, i.e. the austenite is not decomposed into ferrite and carbides at the usual cooling rates of a weld, but transforms to martensite between the M_s and M_f temperatures. Consequently, the following structures can be found in a weld metal cooled to a temperature T (for example, room temperature):

– Martensite alone, or martensite with delta ferrite, if the temperature T is below the M_f point. For instance, this is the case with steels containing 13 % chromium and more than 0·08 % carbon, cooled to normal ambient temperature.

– Martensite or martensite with delta ferrite, and untransformed austenite, if the temperature T lies between the M_s and M_f points. All other things remaining equal, the probability of the presence of untransformed austenite increases when the contents of all alloying elements increase, and it can be necessary in some cases to cool the weld below room temperature in order to obtain a fully transformed structure. By locally enriching the metal in alloying elements, segregation during solidification can also cause the retention of untransformed austenite.

All this is also applicable to the heat-affected zone (HAZ) which is sufficiently near the weld metal to have been raised into the austenite range or beyond it. In practice, depending upon the maximum temperature attained and the duration of the welding cycle, it is possible to find in this zone all the intermediate structural states between that of the unaffected base metal (tempered or annealed) and that of the weld metal.

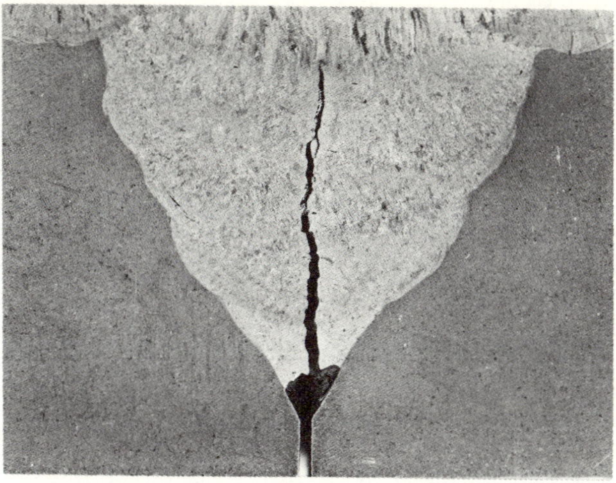

Fig. 4.4. Cracked weld in a 13 % chromium martensitic steel (× 4).

3 Cracking due to martensite formation

For the weld metal and HAZ, the effects of martensite formation are the same as the effects in low alloy steels.

3.1 Mechanical effects

Martensite is hard and brittle, and very notch-sensitive. Its formation is accompanied by a significant expansion of up to 1 % in volume, and in conjunction with shrinkage and thermal gradient effects, this causes a sharp increase in the level of internal stresses in the assembly at a temperature range over which the capacity for plastic deformation is low. There is, therefore, a considerable risk of brittle fracture or cracking.

Cracks are initiated by physical defects in the assembly, which act as stress concentrations or notches, and they can subsequently propagate very rapidly through the whole weld. Lack of penetration or side-wall fusion defects, undercutting or non-metallic inclusions can act as initiation sites for cracking. Fig. 4.4 shows a crack in a weld in a 13 % chromium martensitic steel, initiated at a fusion defect at the root of the joint.

3.2 The role of hydrogen

A further effect, related to hydrogen solubility and diffusion, can contribute to cracking in martensitic steel welds, although 'fish-eyes' are not observed, in contrast to the case with ordinary steels. It was pointed out in chapter 3, § 1.2, that the solubility of hydrogen in ferrite or austenite becomes lower as the temperature decreases, and that the transformation of the austenitic structure into ferrite or martensite is accompanied by an abrupt decrease in the hydrogen solubility. When the austenite contains hydrogen, the martensitic transformation can cause an abrupt supersaturation which results in the liberation of molecular hydrogen at microdefects in the weld (porosity, small bonding defects, non-metallic inclusions). It has been estimated [3] that, with the amount of hydrogen normally present, this liberation in molecular form could take place at room temperature under the pressure of many thousands of atmospheres. Such pressures are liable to initiate brittle fractures and, as this phenomenon depends upon the diffusion rate of hydrogen in the metal, fracture can occur many hours after the welding operation. This type of cracking related to the martensitic transformation arises at a relatively low temperature. It is therefore often called 'cold cracking', as opposed to the cracking of austenitic steels at high temperature which has quite a different mechanism.

3.3 The mechanism of cracking

Cracking due to the mechanical effects of martensite formation and to the presence of hydrogen has been studied in detail with respect to hardenable low alloy steels, as indicated in refs. [4] and [5], for instance. These investigations have enabled the mechanical role of the martensitic transformation to be distinguished, and have explained the localisation of the cracks encountered in practice.

There has been little information published relating to martensitic stainless steels, but from the experience acquired, it can be assumed that the phenomena are similar to those encountered in the case of low alloy steels.

The general characteristics are:

– The location of cracking caused by the mechanical effect of quenching depends on the progress of the martensitic transformation. When the weld metal and HAZ are equally hardenable, transformation normally occurs first in the HAZ and progresses towards the weld metal as a result of the heat distribution resulting from welding. This has a restraining effect, causing preferential weld metal cracking. If, because of differences in chemical composition between the base metal and weld metal, the latter transforms first, cracking takes place preferentially in the base material. The avoidance of this cracking by preheat and post-heat will be considered below.

– The location of hydrogen cracking results from segregation associated with the martensite transformation. The solubility of hydrogen is lower in martensite than in austenite, and thus as transformation takes place, hydrogen is concentrated in austenitic regions remaining untransformed. When these regions finally transform, the resultant martensite is supersaturated with hydrogen, and brittle cracking can occur. Such an effect is partially responsible for the preferential cracking of weld metal when this is last to transform. This effect is complicated by the fact that when hydrogen is present, it is generally introduced by the weld metal. During cooling, this gas diffuses rapidly towards the HAZ, whereas martensite formation progresses in the opposite direction. This explains the formation of 'underbead cracking' in the HAZ close to the fusion boundary, in which it has been shown that hydrogen plays a major role.

Cracking due to hydrogen can arise with no restraint; however, it seems to be facilitated by internal stresses. Both preheat and post-heat of welded joints have an effect on the formation of cracking.

3.4 Limiting carbon content

Other factors remaining constant, the risk of cracking increases with increasing martensite hardness. As in the case of low alloy steels, the hardness of martensitic stainless steels rises rapidly with carbon content. The effect of carbon content on the hardness of 12 % Cr steels quenched from 1050 °C is given below [6]:

Carbon content (wt %)	0·068	0·133	0·206
Hardness (HV)	364	462	480

After quenching from 1050 °C, 12 % Cr steels with 0·45 and 0·60 % carbon show hardnesses of about 580 and 620 HV respectively. Further, hardness generally increases with increasing quenching temperature, and this adverse effect may be particularly significant with respect to metal deposited by welding.

Because of the high hardnesses developed, martensitic stainless steels with carbon contents above 0·25–0·3 % are not normally welded. Above this level, precautions taken to avoid cracking may not give sufficient security.

4 Precautions for the avoidance of cracking

Apart from control of steel carbon content and reduction in joint restraint, additional precautions are necessary when welding martensitic steels. The efficacy of these is dependent on the factors considered above.

4.1 Protection of the weld metal against hydrogen

The effect of hydrogen in inducing cracking precludes arc welding under $Ar–H_2$ mixtures, as well as atomic hydrogen welding. For practical purposes, gas-shielded welding requires the use of pure argon, or slightly oxidising gas mixtures.

Basic coatings are recommended for manual metal-arc welding, since they give deposits with a lower hydrogen content than do rutile coverings.

It is also advisable to avoid any trace of moisture in the welding atmosphere, as may arise for example from condensation, or could be introduced by the fluxes or electrode coatings; the latter in particular must be dried immediately prior to use, at the maximum temperature that can be used without prejudicing the integrity of the coating.

4.2 Preheating of the assembly

Preheating of the joint is the most commonly employed technique for reducing the effects of the martensitic transformation. The joint is heated to a temperature which depends upon various factors, and maintained at this temperature during the whole welding operation, in order to attain the following objectives:

– A reduction in the shrinkage stresses by maintaining the assembly at a temperature at which its elastic limit is lower than at room temperature, thus giving more rapid relaxation of internal stresses. Such stress relaxation becomes more effective as the temperature is higher.

– A reduction in the notch sensitivity of the weld, thus diminishing the risk of brittle fracture. In general, the brittleness of martensitic steels increases as the temperature decreases, and for given conditions of composition, structure and stresses, it is possible to define a temperature below which a crack is likely to propagate rapidly in the metal. These temperatures are not well known for martensitic stainless steels in the as-welded condition, but a rise in the temperature of the assembly always reduces the risk of brittle crack propagation.

– To allow austenite transformation to take place at relatively high temperatures by reducing the cooling rate. This objective is rarely achieved, because it would be necessary to preheat above 400 °C in the majority of cases, in view of the high hardenability of martensitic stainless steels.

– To avoid or limit transformation to martensite of the weld metal and HAZ throughout the welding operation.

To avoid martensite formation, it is necessary that, allowing for the heat input of the welding operation itself, the preheating temperature is high enough for the metal to remain at above the M_s temperature for a sufficiently long time. This is rarely achieved during the welding of chromium–carbon steels, for which the M_s point is relatively high; nevertheless it is possible, especially in the case of martensitic steels with a high content of alloying elements which have the effect of lowering the M_s and M_f temperatures.

However, experience has shown that cracking can be avoided when the preheating temperature allows the weld simply to remain for a sufficient time at a temperature below that of the M_s and above room temperature, i.e. to have a sufficiently slow cooling rate in this range. This has been demonstrated with both low alloy [5] and stainless steels. It should therefore be appreciated that this is really a question of post-heating, and that preheating is not strictly necessary. Nonetheless, preheating gives other advantages mentioned above, and in practice, it is usually difficult to carry out such post-heating without a prior heating of the workpieces.

Besides diminishing the notch sensitivity, the beneficial effect of maintaining the weld between the M_s and room temperature can be explained in the following ways:
– Holding the weld above room temperature undoubtedly facilitates stress relaxation, even for a fully martensitic structure; moreover, it reduces the level of thermal stresses by favouring temperature equalisation.
– The M_f point of martensitic stainless steels is relatively low, and even at a moderate temperature, for example 150–200 °C, small islands of untransformed austenite are retained. These enable hydrogen to remain in solution, and facilitate the relaxation of internal stresses owing to their large capacity for plastic deformation. In this respect, it might be noted that the presence of delta ferrite can also be a favourable factor from the point of view of relief of internal stresses.
– The behaviour of hydrogen in the steel is greatly influenced by a slight rise in temperature. Thus [3] when the temperature rises from room temperature to 200 °C, the pressure of the molecular hydrogen, which is present in microvoids of the metal in equilibrium with dissolved hydrogen, is reduced by a factor of 100 or more. Further, the diffusion rate of hydrogen is increased by a factor of at least 400. It will be appreciated, therefore, that holding between 150 and 200 °C avoids the formation of high pressures, gives a uniform hydrogen concentration, and allows rapid diffusion of the gas out of the weld.

Thus, it is evident that the temperature to which it is necessary to preheat an assembly to avoid cold cracking depends upon several factors, among which are:
– Compositional factors which determine the M_s and M_f temperatures, and have an effect upon the hardness of the martensite, and also upon the presence of delta ferrite.
– Geometrical and technological factors such as the welding process, current, joint restraint and thickness, which affect both the intensity of internal stresses and the cooling rate of the weld.

Consequently it is impossible to define the preheating temperature in advance with accuracy. Nevertheless, depending on local practical conditions, preheating temperatures generally lie between 100 and 300 °C, although 13 % chromium steels containing no more than 0·10–0·12 % carbon can often be welded without preheating if the thickness does not exceed 10 mm (0·4 in.).

When the carbon contents of martensitic steels are relatively high, it is advisable to carry out tempering after welding, as will be seen in § 5 below. In general, the situation can be summarised as [7]:

$C < 0{\cdot}1\,\%$: neither preheating, nor tempering,
$0{\cdot}1\,\% < C < 0{\cdot}20\,\%$: preheating at 250 °C, slow cooling,
$0{\cdot}2\,\% < C < 0{\cdot}5\,\%$: preheating at 250 °C, tempering after welding.

As the necessity for preheat often constitutes a practical problem, especially with higher preheating temperatures, it is always recommended that the minimum permissible temperature be ascertained by carrying out a procedural test which is as representative of practice as possible.

Remarks. The martensitic 16–20 % chromium steels containing less than 0·10 % carbon with the addition of nickel can very often be welded without preheating. In the as-welded condition, these steels contain a high proportion of delta ferrite, while the low carbon martensite which they form is relatively ductile at room temperature. Both these factors significantly reduce the risk of cold cracking.

4.3 Use of an austenitic filler metal

Except for the case of overlaying of very massive components, the use of a chromium–nickel austenitic filler metal allows martensitic steels to be welded without preheat. As the weld metal does not harden and remains very ductile at room temperature, cracking due to mechanical effects in the martensitic zone of the base metal is avoided. Moreover, since hydrogen is very soluble in austenite, and diffuses slowly in this phase, underbead cracking in the base metal is also avoided.

In the chapter dealing with heterogeneous assemblies (chapter 9) the choice of a filler metal will be reconsidered with particular regard to the need for heat treatment after welding, and the problems arising from the heterogeneity of the welded joint.

5 Heat treatment after welding

Depending upon the temperature at which the welding operation finishes and the precautions taken to avoid cracking, the weld can consist of martensite or a mixture of martensite and untransformed austenite. The amount of austenite present immediately on completion of welding is greater at temperatures closer to the M_s, and cooling to room temperature before the assembly is put into service causes transformation to a fully martensitic structure.

Thus the weld must be subjected to a tempering operation† for some hours at a temperature generally between 600 and 850 °C to obtain properties of strength and ductility comparable to those of the base metal. From the point of view of the mechanical characteristics, tempering does not give the same results when applied to a purely martensitic structure as when applied to a structure having a high proportion of austenite. In the first case, the martensite decomposition gives rise to a structure composed of ferrite and uniformly distributed carbides, enabling the best compro-

† The weld is rarely subjected to annealing as the maximum softening is rarely required, and because annealing is carried out at a high temperature, which is often practically difficult to achieve.

mise to be obtained between strength and ductility. In the second case, the austenite decomposition leads to a ferritic structure in which carbides precipitate preferentially at the grain boundaries; in view of the relatively coarse grain size of the as-welded structures, this leads to a rather brittle structure.

Consequently, if the welding operation is terminated at close to the M_s point and the assembly is immediately placed in the tempering furnace, the optimum mechanical properties are not achieved, even though the procedure is completely reliable from the point of view of avoiding cracking. The best properties are obtained either by subjecting the assembly to a complete quenching and tempering treatment when this is possible, which entirely eliminates the structural consequences of the welding operation, or by carrying out the heat treatment as illustrated by the following example.

A martensitic creep-resisting steel is considered, having a composition of 12 % Cr/0·2 % C/1 % Mo/0·4 % W/0·3 % V. After oil quenching from 1050 °C and tempering at 750 °C for four hours, this steel has a tensile strength of 780 N/mm^2 (50 tonf/in.^{2}), and a UF impact value of 60 J/100 mm^2.

Welds were made in highly restrained heat-treated plates of 15 mm thickness, using basic electrodes of matching composition to the base metal. Under these conditions, the preheating temperatures necessary to avoid cracking during a natural cooling of the weld is 100–150 °C. However, the assemblies have been preheated at a higher temperature, in order that the temperature at the end of welding is above 260 °C, the M_s point of the alloy. The structure on completion of welding is thus mainly composed of unstable austenite.

After welding, assembly A was maintained for 1 h at 300 °C, then subjected straight away to tempering at 750 °C for 4 h; assembly B was cooled at 150 °C, kept for 1 h at this temperature and then subjected to the same tempering as assembly A. The corresponding thermal cycles are shown diagrammatically in fig. 4.5.

After these treatments, the mechanical characteristics of the welds are as follows:

	Tensile strength [N/mm^2 (tonf/in.^{2})]	UF impact value [J/100 mm^2]
Weld A (tempering carried out on austenite)	650 (42)	below 10
Weld B (tempering carried out on martensite)	750 (48)	55

The microstructures of the weld metals and HAZ are shown in fig. 4.6. The structure of weld A is ferritic with agglomerated carbides, whereas weld B has the characteristic structure of tempered martensite. The pre-

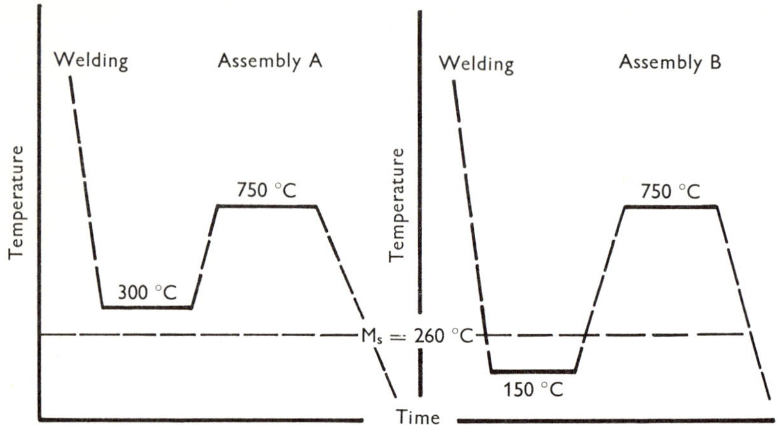

Fig. 4.5. Thermal cycles after welding a martensitic steel containing 12 % chromium, 0·2 % carbon, 1 % molybdenum, with additions of tungsten and vanadium.

sence of delta ferrite is also apparent, clearly visible as the light etching phase in the weld metal of weld B; the fact that weld B has slightly lower properties than those of the parent material must be attributed to the presence of this phase. This difference would be eliminated if a complete heat treatment were applied, as quenching from 1050 °C prior to tempering considerably reduces the quantity of delta ferrite.

This procedure allows welds to be obtained having mechanical characteristics similar to those of the base metal, but requires a knowledge of the M_s temperature of the steel under consideration. With massive components, it necessitates careful control of the temperature of the assembly, because the risk of cracking increases when the temperature to which the weld is allowed to cool decreases. Thus, it is still advisable to perform a procedural test to define conditions ensuring reliability from the cracking point of view.

Remarks. When treatments of the types A and B described above are applied after welding, the formation of secondary martensite can cause a certain brittleness of the assembly. This phenomenon, which will be considered again in the case of age-hardening stainless steels (chapter 8, §3.3), can be explained on the following basis. The chemical heterogeneity of the weld metal can prevent the transformation of some austenite islands into ferrite on tempering, and they act only as sites for precipitation of carbides. Hence, the austenite becomes impoverished in carbon and can transform to martensite during cooling after tempering, with an associated embrittling effect on the weld. The heat treatment is said to 'destabilise' the austenite.

– From the point of view of relief of internal stress and uniform distribution of hydrogen, treatment at a temperature lower than the normal tempering temperatures is sufficient. When applied to a fully martensitic

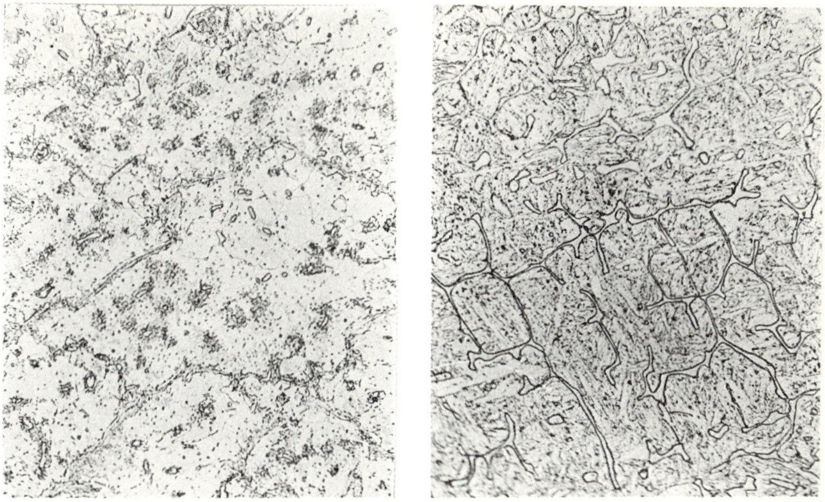

Cast metal

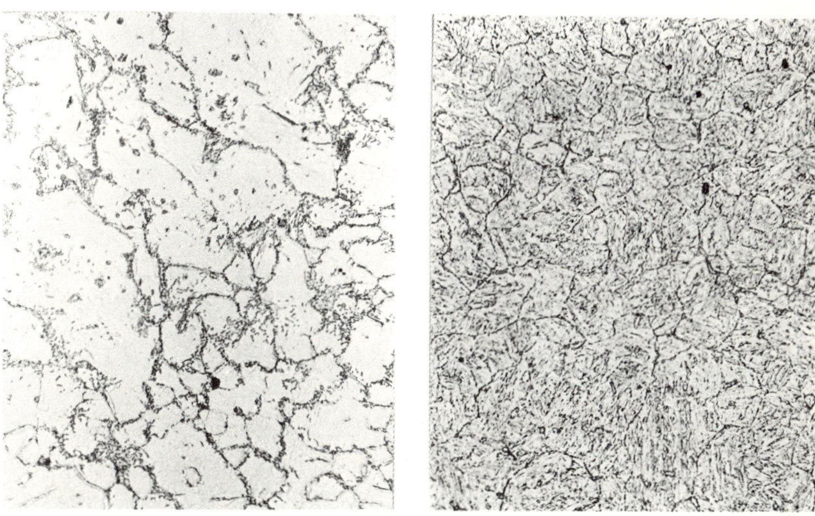

Zone affected by the heat

Fig. 4.6. Effect of the thermal cycle after welding upon the micro-structure of welds A and B in fig. 4.5 (× 500).

structure, treatment at about 300 °C slightly improves the ductility. These treatments at low temperature are generally known as 'relieving treatments', or 'low temperature heat treatments'.

– It has already been pointed out that the use of martensitic steels with a low carbon content and an addition of nickel often permits the avoidance of preheating. Weldments having this composition can be quite ductile in the as-welded condition, because the low carbon martensite that is formed is not very hard. The situation is further improved when the weld is made using a number of small runs, because the deposition of successive runs subjects the underlaying deposited metal to a short tempering. For example, a weld in 15 mm (0·6 in.) thick steel containing 16·5 % chromium, 0·04 % carbon and 5 % nickel, performed in eight runs, has a hardness of 400 HV in the final runs, and a hardness of 320 HV in the first run.

This permits, in some cases, the avoidance of tempering after welding.

– The difficulties arising from the thermal precautions necessary to weld air-hardening steels should not be exaggerated. These precautions are now well recognised, and are regarded as commonplace because of the more and more extensive use of martensitic stainless and low alloy high-strength steels. Moreover, suitable equipment has become available which facilitates the welding of these steels under controlled thermal conditions. If appropriate thermal precautions are taken and weldable grades of steel are selected, these steels can be welded with little risk of cracking.

6 Application of the various welding processes to martensitic stainless steels

All the conventional fusion processes are applicable to martensitic stainless steels, provided that processes or consumables giving high hydrogen contents in the deposited metal are avoided, as mentioned above.

Resistance welding of sheet, together with resistance or flash butt welding of bars may also be carried out without difficulty, on condition that the welding cycle includes a tempering or stress relaxation operation. When such treatment cannot be performed on the welding machine, it is necessary to allow for a complete or local heat treatment of the welded assembly.

The necessity for heat treatment constitutes a practical drawback to rapid processes like the resistance welding methods. For this reason, whenever the high mechanical properties of martensitic steels are not essential, it is preferable to employ semi-ferritic 13 % Cr steels, with very low carbon contents, or ferritic steels with aluminium additions as will be considered in chapter 5.

References

[1] R. Tricot and R. Castro. Mémoires Scientifiques, Rev. Mét. **63** (1966) 657.
[2] R. Castro. Comptes Rendus Académie des Sciences (16 Oct. 1961) 1690.
[3] C. E. Sims. *Gases in Metals*, A.S.M. (1953).
[4] P. Bastien. Conférence Houdremont, 1959. *Soudage Techniques connexes*, **13** (1959) no. 9–10.
[5] H. Granjon. Adams Lecture, 1961. *Soudage Techniques connexes*, **15** (1961) 245.
[6] K. J. Irvine, D. J. Crowe and F. B. Pickering. *J. Iron Steel Inst.* **195** (1960) 386.
[7] G. E. Linnert. *Metals Eng. Quart.* (Nov. 1967) 1.

5

WELDING OF STAINLESS AND HEAT-RESISTING SEMI-FERRITIC AND FERRITIC STEELS

1 General comments – composition and structure

1.1 Semi-ferritic stainless steels

It was shown in chapter 1 that stainless and heat-resisting semi-ferritic steels are iron–chromium-based alloys with a composition such that, on cooling, they pass through a temperature range over which their structure is composed of delta ferrite and austenite. The latter transforms to martensite on rapid cooling, whereas the delta ferrite remains unchanged.

Examination of the vertical sections of the equilibrium diagram for iron–chromium industrial alloys containing 0·05 %, 0·10 % and 0·20 % carbon, described in chapter 4, §1 (fig. 4.1), shows that this mixed structure can be obtained with various combinations of chromium and carbon contents, taking into account the presence of the usual minor alloying elements. To obtain a semi-ferritic steel, it is necessary to relate the carbon content to a minimum proportion of chromium, this proportion being higher as the carbon content increases.

The most common semi-ferritic steels in practice are the 17 % chromium grades containing 16–18 % of this element and 0·05–0·12 % carbon. They have the following characteristics, as compared to 13 % chromium martensitic steels:

- They have a better corrosion resistance.
- They are hardened only moderately by quenching.
- As a result of their structure and low carbon content, they are quite amenable to deep-drawing and cold-forming.

Figure 5.1 shows the vertical section of the equilibrium diagram for commercial 17 % chromium alloys containing the common minor alloying elements[1]. For the carbon contents under consideration, the alloy solidifies as delta ferrite, then enters the temperature range over which austenite appears, and then that over which chromium and iron carbides are formed. When cooling proceeds at a sufficiently slow rate, the austenite changes into alpha ferrite, so that below 850–900 °C, the alloy is composed of (delta and alpha) ferrite and carbides. With rapid cooling, the austenite

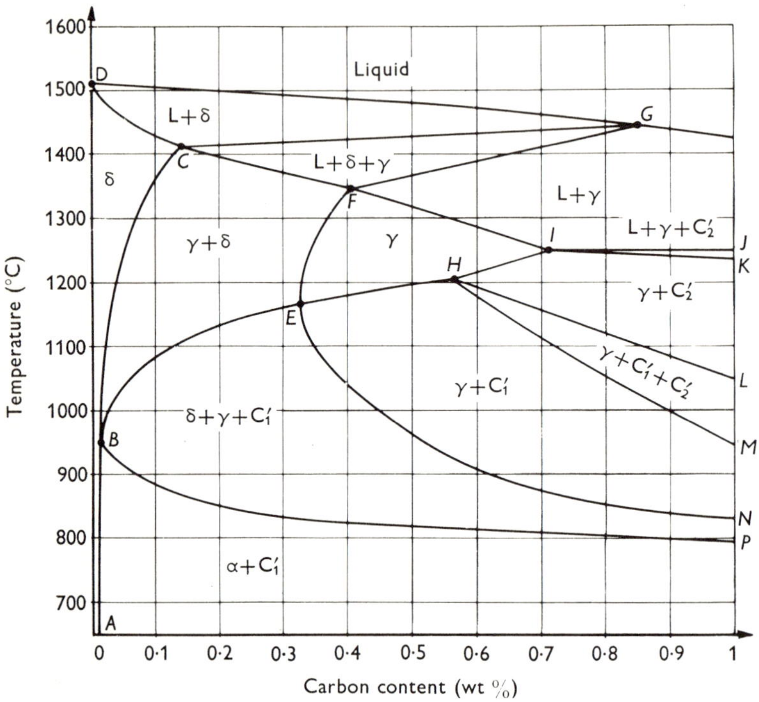

Fig. 5.1. Vertical section at 17 % chromium of the phase diagram of an industrial Fe–Cr–C alloy (Ni ⩽ 0·13 %/N ⩽ 0·013 %/Si ~ 0·4 %/Mn ~ 0·5 %).

is transformed into martensite, and the alloy then consists of ferrite, martensite and carbides.

When these steels are raised to a sufficiently high temperature for their structure to be purely ferritic (above about 1350 °C), they suffer very rapid grain growth. Although austenite appears on cooling in increasing amounts down to approximately 1100 °C, it is formed from delta ferrite, and its eventual disappearance is caused by regression into alpha ferrite. This mechanism does not modify the grain boundaries of the ferritic structure as a whole. Thus, the structure formed at very high temperature, especially the coarse grain size, cannot be removed by a grain-refining heat treatment, as is possible by the alpha–gamma transformation in low alloy steels. In order to eliminate the coarse structure, it is necessary to apply sufficient plastic deformation to the metal to induce recrystallisation to a fine grain size on subsequent reheating. This effect is very important from the welding point of view.

The normal hot-working temperatures for this type of steel lie mainly in the two-phase range, although in the form of forgings and plates, these steels are always heat treated at about 800 °C prior to use, giving a structure of ferrite and carbides. The extent of the two-phase range depends greatly on the minor and other alloying elements. Both carbon and nitrogen have the effect of widening the austenite range towards high temperatures, or, at a fixed temperature, of reducing the amount of ferrite in the two-phase range. Nickel has a similar influence, and it was pointed out in chapter 4, §1, that an addition of 4 % of this element to a 17 % chromium steel transforms it into a martensitic steel. Silicon and certain alloying elements like aluminium, molybdenum, titanium or niobium, have the reverse effect.

A number of types of precipitate containing carbon and nitrogen are also formed in these steels [2], but these will not be considered further as, at present, they do not seem to play a leading part in the welding of 17 % chromium steels.

It has been found that welds in 17 % chromium steel may suffer cracking as a result of the formation of martensite. Various techniques have been employed to minimise or avoid this difficulty, since together with other problems, it has for a long time seriously hindered the use of this economical type of steel for welded constructions.

The first method consists of adding a certain quantity of 'ferrite formers' to the base material, rendering it ferritic at all temperatures, and avoiding the formation of austenite which subsequently transforms to martensite during cooling. This can, for example, be achieved by adding about 0·7 % titanium or 1·2 % niobium to the initial composition. However, these steels have an excessive tendency for grain growth to occur in the HAZ immediately adjacent to the weld bead, while some cracking susceptibility persists as a result of the concentration of impurities in the limited grain boundary surface area. Some authors [3] have suggested that these additions of titanium or niobium should be associated with small amounts of other elements, so that dispersions of phases are formed which oppose the grain coarsening. In the case of titanium, it has been proposed that aluminium be added at the same time, causing the formation of TiO_2, TiN, Al_2O_3, AlN, etc.

The second method [4] is based on modifying the analysis of the 17 % chromium steel in such a way that the austenite formed at high temperature in this semi-ferritic steel is sufficiently stable to remain untransformed to martensite at low temperature. For example, this can be obtained by adding about 5 % manganese and 1 % silicon, plus various minor additions. After welding, the structure of the HAZ and weld bead is austenitic–ferritic, and thus little embrittled.

It must be pointed out that, if the conditions of welding are such that moisture is present during the operation (residual moisture in the electrode coating, for instance), hydrogen can be introduced into the metal and lead to cracking which generally takes place some time after welding. If such hydrogen pick-up is suspected, it is advisable to carry out tempering after welding, permitting hydrogen to diffuse out, and simultaneously increasing the ductility of the metal. The temperature and duration of this tempering will, of course, be related to the properties required from the material.

Also included in the category of semi-ferritic steels are 13 % chromium alloys in which the carbon content is below about 0·07 %, i.e. low enough for a two-phase range to exist, as shown by fig. 4.2 of chapter 4. These steels may be preferred to 13 % chromium martensitic steels when an alloy is required that hardens only moderately on quenching, can be easily cold-worked, and is weldable in light section sizes without risk of cracking. Their metallurgy is analogous to that of 17 % chromium steels, although with a reduced tendency to grain growth.

1.2 Ferritic stainless and refractory steels

Ferritic stainless steels have a composition close to that permitting the alloy to have a structure composed only of ferrite and carbides at all temperatures. In industrial practice, they are represented by steels containing 25–30 % chromium and up to about 0·3 % carbon. From fig. 4.1 it can be seen that, within these limits, compositions which are relatively poor in chromium and rich in carbon can still pass through the two-phase range, and consequently have a certain capacity for hardening.

Ferritic stainless steels are always employed after annealing at 800 or 900 °C, in the ferrite+carbides range. Like 17 % chromium semi-ferritic steels, they are very sensitive to grain growth at elevated temperature.

25 % chromium steels generally possess better corrosion resistance than 17 % chromium alloys. However, they are used mainly because of their good resistance to oxidation and sulphidation at high temperature, and are generally refractory steels. Their application in welded constructions is quite limited on account of the brittleness of the joints obtained, as considered below. However, 26 % Cr/1 % Mo steels have recently become commercially available with extremely low carbon and nitrogen contents ($C+N < 0{\cdot}010$–$0{\cdot}015$ %), and are reported to be weldable without marked embrittlement, especially if the usual minor elements and certain impurities are also maintained at very low levels ($Si < 0{\cdot}1$ %, $Mn < 0{\cdot}2$ %, $O_2 < 0{\cdot}02$ %) [5].

Finally, the ferritising effect of aluminium is turned to account to obtain steels which, though they contain about 13 % chromium, less than 0·07 % carbon and nearly 0·25 % aluminium, are nevertheless almost entirely ferritic. They have the advantage of being welded without difficulty, are easily cold-worked, and are not hardened by quenching.

In the course of this chapter, particular attention will be paid to semi-ferritic steels, which are by far the most widely employed.

2 Microstructural changes on welding

2.1 17 % chromium steels

To describe the structural transformations caused by the welding operation, two regions must be distinguished within the weld:

– A region A of the HAZ in which the maximum temperature attained corresponds to the ferrite–austenite range of the equilibrium diagram.

– A region B composed of the remaining HAZ and of the weld metal, in which the metal has reached the fully delta ferrite range.

Figure 5.2 gives the percentage of austenite present at equilibrium in a 17·3 % chromium, 0·09 % carbon, 0·013 % nitrogen steel. It is apparent that the austenite proportion reaches a maximum around 1100 °C, and decreases as the temperature rises towards 1330 °C. Fig. 5.3 is a metallographic section of a fusion weld made in this type of steel, on which the regions A and B have been defined.

In region A, the reheating has caused the formation of an austenite rather similar to the austenite under equilibrium conditions, in the form of relatively small and well-distributed bands; this austenite can be slightly depleted in chromium (up to 2 %) as compared to the ferrite. On fast cooling, reversion to alpha ferrite is largely avoided, and the austenite changes into martensite, appearing as the dark etching areas in fig. 5.3 (region A). Little grain growth takes place in this region.

In region B, the metal has reached the pure delta ferrite range, and the grain size has increased considerably. During the rapid passage through the two-phase range on cooling, a marked quantity of austenite has been formed at the boundaries and inside the grains in a distinctive acicular† form resembling the Widmanstätten structure of low or non-alloyed carbon steels. This is an unstable austenite which is formed solely by fast cooling from the delta ferrite range, and has a composition little different from the original ferrite. The quantity of austenite which appears under these conditions depends upon the alloy composition, in the same way as does the austenite content at equilibrium; it also depends upon the cooling rate, but it is impossible to avoid its formation entirely, even by the most rapid water quenching. The austenite formed in region B is then transformed into martensite as the weld cools to room temperature.

It will be noted that there is no marked structural difference between the weld metal and the portion of HAZ raised into the fully delta ferrite range. It is, however, possible to distinguish between these regions, because the metal of the HAZ retains an equiaxed grain structure, whereas the weld metal possesses a columnar structure resulting from the directional solidification.

† In fact, these needles are sections through plates.

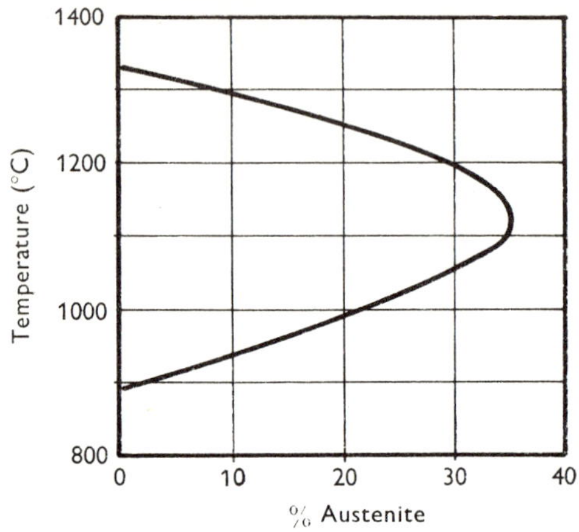

Fig. 5.2. Proportion of austenite existing at equilibrium in a steel containing 17·3 % chromium, 0·09 % carbon and 0·013 % nitrogen.

2.2 Low carbon 13 % chromium steels

Similar phenomena occur during the welding of low carbon 13 % chromium steels. However, in the usual range of composition, the grain growth is often less pronounced than in the case of 17 % chromium steels, because the transition into the purely ferritic region takes place at a higher temperature: other things remaining constant, less time is spent in the temperature range where the grain growth is very fast.

Semi-ferritic 13 % chromium steels generally have a composition such that their welds contain more martensite than those of 17 % chromium alloys (see § 5).

3 Mechanical consequences of fusion welding

The mechanical properties of a fusion-welded joint result from the presence of a coarse structure and from the nature of the physico-chemical constituents which are formed on welding.

In general, the high chromium ferrite which is formed at elevated temperatures in alloys of normal industrial purity is very notch sensitive. This sensitivity is particularly apparent in metal heated into the fully ferritic range, as evidenced by conventional toughness testing. As a result, the material is very brittle on impact loading, this being more pronounced as the notch is made more acute. The martensite which is formed in the weld metal and heat-affected zones accentuates this brittleness even further.

As an example, the Charpy V impact value of a weld in a 17 % chromium steel is below 5 J/100 mm^2 (3 ft lb), but it can reach 80 J/100 mm^2 (54 ft lb) when the notch has a radius of 5 mm (0·2 in.).

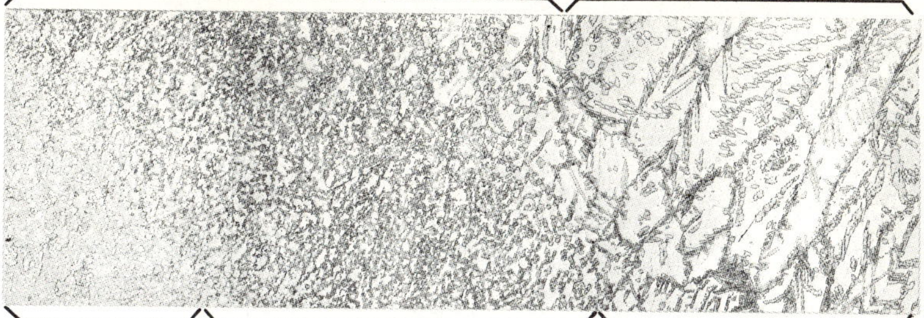

Fig. 5.3. Section of a fusion weld in a 17 % chromium steel (× 35).

This notch sensitivity is not accompanied by a high hardness, as is the case with martensitic steels; moreover, transformation to martensite takes place to only a limited extent, so that problems associated with hydrogen diffusion are avoided. The result is that semi-ferritic steels, and especially ferritic steels, are insensitive to the cracking which takes place in martensitic alloys. In fact, the weldability of these steels is good, and sound welds can be made in quite thick material without preheating. It is shown below that the notch sensitivity of fusion-welded semi-ferritic and ferritic steels disappears as the temperature rises, and this has sometimes led to recommendations that the workpieces be preheated to a sufficient temperature so that this sensitivity does not exist. However, this precaution does not seem to be of much practical use, and it has the disadvantage of promoting grain growth by slowing down the rate of cooling.

4 Methods of reducing the notch sensitivity of fusion welds in 17 % chromium steel

4.1 Martensite tempering

The presence of martensite is a cause of brittleness in the true sense of the word; it reduces the capacity for slow plastic deformation in the absence of notches (bend tests, for instance) as well as the toughness. Hence, it is advisable to heat-treat the assembly at about 800 °C, which transforms the martensite into alpha ferrite and carbides; carbide precipitation also takes place in the delta ferrite, slightly improving its plastic properties. Fig. 5.4 shows the weld metal of a weld in 17 % chromium steel which has been subjected to this treatment.

Treatment at 800 °C slightly increases the toughness of welded joints. For example, the Charpy V impact strength of specimens taken from a 20 mm (0·8 in.) thick weld does not exceed 3 J/100 mm² (2 ft lb) in the as-welded condition, whereas it can reach 10 J/100 mm² (6 ft lb) after heat treatment. However, the minimum values remain low, and are extremely variable

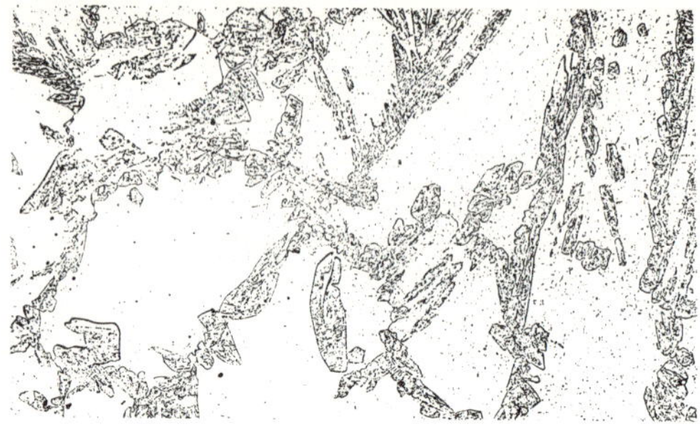

Fig. 5.4. Weld metal of a weld in a 17 % chromium steel after tempering at 800 °C (×180).

according to the notch direction with respect to the orientation of the solidification structure.

Heat treatment at 800 °C gives the weld good plastic properties during slow deformation of unnotched testpieces, and welds can be bent through 180 ° round formers of diameter twice the material thickness without cracking. Thus, the low toughness clearly illustrates a notch sensitivity, and not intrinsic brittleness. Generally, fusion welds heat treated at 800 °C have mechanical properties which, with the exception of impact strength, are similar to those of the unaffected base metal.

This notch sensitivity is detrimental only when the weldment is liable to be subjected to violent shocks, and when the joint configuration leads to the formation of stress concentrations.

Finally, treatment at 800 °C gives a general relief of welding stresses, and in particular removes any local concentration of stress which can have an effect similar to that of a notch.

4.2 Use of welds at above room temperature

In general, toughness tests carried out on steels having a martensitic or ferritic structure show that this characteristic is a function of the temperature; for every steel, there exists a temperature below which the notch toughness of the metal is very low (brittle fracture), while at some higher temperature, the notch toughness is increased (ductile fracture). In the intermediate (or transition) range, the impact values show considerable scatter, but their average value increases rapidly as the temperature rises. It is possible to define in various ways a temperature known as the toughness transition temperature, which represents the boundary between the low toughness range and the high toughness range; this temperature is never very far from the middle of the transition range.

The chromium-rich ferrites found in fusion-welded joints have low toughness because room temperature is below their transition temperature. Fig. 5.5 shows the variation with respect to temperature of the impact

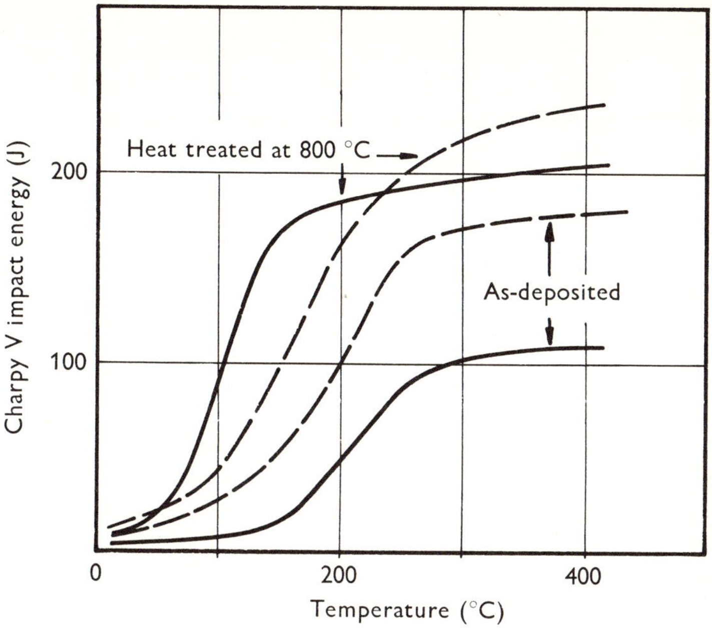

Fig. 5.5. Variation of weld metal Charpy V impact values with respect to temperature: —— 17 % chromium steel; – – – steel with 16 % chromium and 0·9 % niobium.

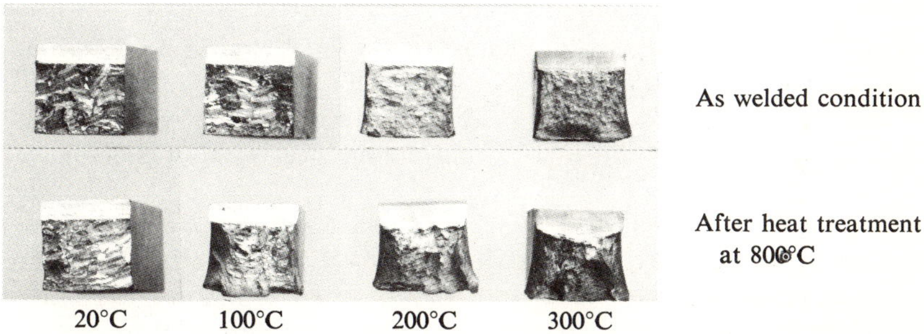

Fig. 5.6. Weld in 17 % chromium steel. Impact test specimens fractured at different temperatures: transition from brittle to ductile fracture.

strength of a 17 % chromium steel weld deposit in the as-welded condition, and after heat treatment at 800 °C (curves in full lines). It can be seen that the middle of the transition range is situated around 100 °C for the heat-treated weld deposit and about 200 °C for the as-deposited weld metal. The former reaches 200 J/100 mm^2 (120 ft lb) at 150 °C, and although the impact strength of the latter does not increase so quickly with increasing temperature, it nevertheless reaches 100 J/100 mm^2 (60 ft lb) at 250 °C. Fig. 5.6 shows the fracture faces of the impact testpieces, and illustrates the transition from a brittle to a ductile type of fracture.

These results show that there is no risk of notch sensitivity in a 17 %

chromium steel weld used at above 100 °C, after heat treatment at 800 °C.

4.3 Compositional modifications to avoid martensite formation

The improvement in impact strength resulting from heat treatment at 800 °C is mainly due to the martensite tempering. A similar result can be obtained without heat treatment, and without increasing the chromium content which would have the effect of raising the notch sensitivity of the ferrite. For this purpose, steels with an addition of about 0·5 % titanium or 1 % niobium are employed, and these alloys allow welds to be obtained which are practically free from martensite in the as-welded condition. The effect of titanium and niobium can be explained in two ways. When dissolved in the metal, both these elements increase the stability of delta ferrite, and at all temperatures decrease the quantity of austenite at equilibrium. Moreover, at very high temperatures, they form very stable carbides and so act in the same way as a decrease in carbon content. Niobium has the advantage over titanium of being less readily oxidised and can be included in the composition of covered electrodes (chapter 3, §4.1).

Figure 5.7 shows the ferritic structure of a weld in a steel containing 16 % chromium, 0·09 % carbon and 0·9 % niobium, in the as-welded condition; the presence can be noted of an eutectic containing niobium, carbon and nitrogen at the boundaries of the solidification structure, together with numerous niobium-rich carbides uniformly distributed throughout the whole structure. In the as-deposited condition, the Charpy V impact strength of this weld is 10 J/100 mm^2 (6 ft lb), a value similar to that obtained for a weld in 17 % chromium steel without niobium after tempering.

Figure 5.5 shows (dashed curves) the variation in impact strength with respect to temperature of a weld in a 16 % Cr/Nb steel both as-welded and after heat treatment at 800 °C. In the as-welded condition, addition of niobium leads to an improvement in the impact strength at all temperatures, the improvement being due mainly to the absence of martensite, and also to the fact that the ferrite contains less carbon in solution. As a further result of these effects there is less difference between the as-welded and heat-treated states. Following heat treatment, the impact transition range is shifted towards lower temperatures, and is wider than is the case for a niobium-free 17 % chromium weld. This is probably due to the presence of niobium in solution in the ferrite.

The elimination of martensite does not, of course, suppress the notch sensitivity of the ferrite at room temperature. Nevertheless, additions of titanium or niobium can be advantageous in the case of fusion welds in thin materials, working under relatively low loads, since these may be put into service without heat treatment.

Fig. 5.7. Martensite-free structure of a weld in a steel containing 16 % chromium, 0·09 % carbon and 0·9 % niobium (×450).

Some other elements can have a similar effect. The most important ones are: aluminium, which allows fully ferritic steels to be obtained containing only 13 % chromium; molybdenum, added to improve the corrosion resistance of 17 % chromium steels; silicon, which increases their oxidation resistance at elevated temperature.

On the other hand, it is possible to transform a semi-ferritic steel containing 17 % chromium and 0·08 % carbon into a martensitic steel by the addition of 2–4 % nickel (chapter 4, § 1). After tempering, welds in this type of material have good strength and ductility characteristics.

4.4 Use of an austenitic filler metal

It is very easy to weld a 17 % chromium steel with the use of an austenitic filler metal, such as one containing 18 % chromium and 10 % nickel.

As will be seen in chapter 6, the weld deposit is insensitive to the presence of notches, and does not require any heat treatment after welding. The metallurgy of this type of weld will be considered further in chapter 9, dealing with dissimilar metal joints (§ 2.4).

Of course, the structure of the HAZ remains unaltered, and the base metal adjacent to the fusion line retains its notch sensitivity. However, the properties of the joint as a whole are improved because these low toughness regions are narrow, and the high capacity for plastic deformation of the weld metal reduces their effects. Fig. 5.8 shows some sections from welds made in 6 mm (0·24 in.) thick plates of a 17 % chromium steel, using an austenitic filler metal. The width of the coarse grain zone con-

MIG using fine spray conditions

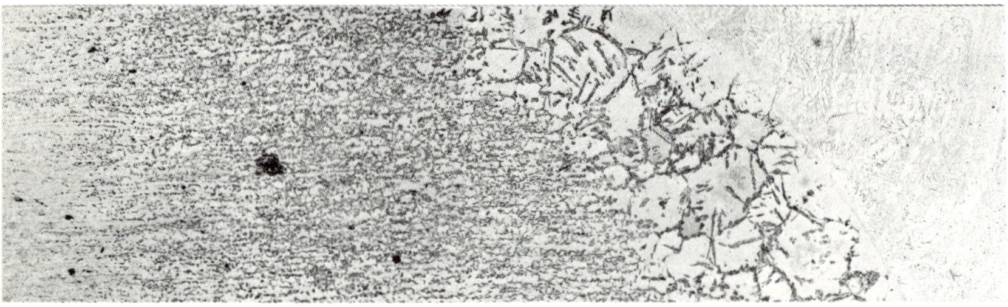

Covered electrode

Fig. 5.8. Sections of welds in a 17 % chromium steel welded with austenitic filler metal (×40).

taining acicular austenite does not exceed 1·5 mm (0·06 in.) in the most unfavourable case, and the small width of the notch-sensitive zone can, in some situations, avoid the necessity for heat treatment after welding.

4.5 Choice of fusion-welding process

The welding process has an effect upon the mechanical properties of 17 % chromium steel welds because it determines the width of the base metal zone heated into the fully ferritic region where grain growth is very rapid. The welding process also influences the cooling rate, and hence affects the time during which grain coarsening can occur, and the amount of austenite formed in the HAZ and weld metal.

On this basis, the manual metal-arc process is preferable to the spray MIG or submerged-arc processes, because it permits the heat input into the assembly to be reduced, for the same amount of deposited metal. The example shown in fig. 5.8 illustrates this fact. In both cases, 6 mm (0·24 in.) thick plates were welded in two passes, with the sections being taken at the junction of the two runs, where the depth of the HAZ is a maximum; the advantage of the covered electrode process is clearly visible. The short circuiting MIG process and, to a lesser extent, the TIG process with filler metal additions, are also advantageous.

In the gas-shielded processes, it is preferable to employ pure argon for TIG welding and argon with 1 % oxygen for MIG welding, for the reasons indicated in chapter 3, §§ 6.2 and 6.3. The presence of nitrogen in the arc atmosphere must be carefully avoided, as this would increase the quantity of martensite in the weld.

Finally, it is necessary to mention that welding tests on 17 % chromium steels have been carried out in which the joint faces and successive welding runs were peened in order to cold-work the metal, and induce its fine recrystallisation during the deposition of further runs. Some limited improvement in toughness could be obtained, but this procedure does not appear to alter significantly the notch sensitivity of welded joints at room temperature.

4.6 Remarks

By employing the procedures and heat treatments described above, sound welds in 17 % chromium steel can readily be obtained with sufficiently high capacity for plastic deformation for practical purposes at room temperature in the absence of notches. However, the importance attached to the concept of toughness considerably restricts the use of these steels for applications such as critical steam generator components, even when lightly stressed, at least in France.

On the other hand, their use for fusion-welded components is continually increasing in other fields. Thus, the fabrication of rolled welded tubes, used for instance in metallic furniture or for piping of relatively small diameter which is not subjected to impact loading, is developing appreciably.

5 Fusion welding of low carbon 13 % chromium steels

What has been said about 17 % chromium steels is applicable to low carbon 13 % chromium semi-ferritic steels, particularly with respect to procedures intended to improve the toughness of welded joints.

However some differences exist, mainly because low carbon 13 % chromium steels generally contain a higher proportion of martensite in the as-welded condition than do 17 % chromium alloys. This is illustrated in fig. 5.9 which shows the microstructure of a weld metal containing 13 % chromium and 0·06 % carbon, with the martensite being visible as the dark etching phase.

In spite of the increased martensite content, this type of steel can be welded in moderate thicknesses without preheating, owing to the low carbon content and the presence of ferrite.

The properties of fusion welds are, in fact, intermediate between those of welds in martensitic and 17 % chromium steels. As in the latter material, the welds have a low toughness, but the martensite plays a major role while the relatively low chromium ferrite is less notch sensitive than the ferrite of 17 % chromium steels. For this reason, tempering after welding

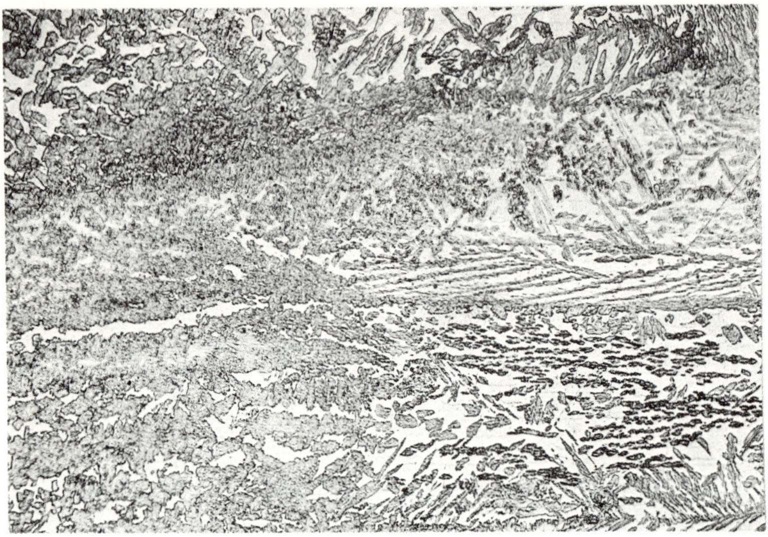

Fig. 5.9. Weld metal of a weld in a steel containing 13 % chromium and 0·06 % carbon (×45).

Heat treatment	Tensile strength [N/mm² (tonf/in.²)]	Elastic limit [N/mm² (tonf/in.²)]	Impact value UF [J/100 mm² (ft lb)]
None	800 (52)	730 (47)	3 (2)
750 °C – 4 h	540 (35)	360 (23)	20 (15)

is quite effective. This is indicated by the table above, giving the mechanical properties of the weld shown in fig. 5.9, made in 20 mm (0·8 in.) thick plate without preheating.

6 Fusion welding of ferritic steels

In §4.3, steels containing 16 or 17 % chromium were considered, which are rendered fully ferritic from the welding point of view by an addition of titanium or niobium. The extent of grain growth in these steels is little different from that in conventional 17 % chromium steels, because it is restrained at high temperature by the presence of stable carbides of titanium or niobium.

The situation is different regarding 13 % chromium steels containing aluminium and 25–30 % chromium steels, since welds in these alloys possess a completely ferritic structure with very coarse grains after cooling. Certain compositions with relatively low chromium and high carbon contents may, however, contain a small amount of martensite in the as-welded condition, formed in a similar way to that in 17 % chromium steels.

Welds in stainless and high temperature ferritic steels are particularly notch sensitive at normal ambient temperature, especially when the chromium content is high, although their hardness remains moderate, and their impact energy transition curve appears similar to that of 17 % chromium steels.

Because of this notch sensitivity, refractory ferritic steels containing 25–30 % chromium are seldom employed in welded fabrications. When welding is carried out, an austenitic or austenitic–ferritic type of filler metal containing 24 % chromium and 12 % nickel, or 29 % chromium and 9 % nickel is normally employed.

It is also uncommon to use 13 % chromium ferritic steel for fusion-welded assemblies. If necessary, fusion welding can be carried out with a 13 % chromium filler metal, or an austenitic consumable containing 18 % chromium and 10 % nickel.

7 Use of resistance welding processes

The principle of these processes was outlined in chapter 3, § 7. They are of interest for the welding of 17 % Cr alloys and semi-ferritic and ferritic steels in general, for the following reasons:

– The volume of metal heated to the temperature range giving rapid grain coarsening is generally quite small. Moreover, the cooling rate is often very high.

– The welding operation is carried out by holding the workpieces together under pressure. This permits the expulsion of molten metal from the joint, and the plastic deformation of the coarse grain regions, thus giving the possibility of developing a fine grain size by recrystallisation.

Examples are given below of the metallurgical effects of the application of these processes to 17 % chromium steels; these effects are similar for the other steels of this group.

7.1 Resistance welding of sheet

After the welding phase, the operation consists of a forging period and a short heat treatment permitting recrystallisation and the beginning of martensite tempering. A high pressure is recommended when welding this type of steel, and when the settings of the machine are suitably selected, welding can be carried out with a very small quantity of molten metal and a very moderate grain growth. The resultant welds have a high capacity for deformation at room temperature, and give no indication of brittleness in service. The right-hand side of fig. 5.10 represents a section through a spot weld made under these conditions. Excessive heating, however, can cause the appearance of the coarse structure characteristic of molten metal, as well as a deep indentation of the plates; this is illustrated in the left-hand side of fig. 5.10. It should, however, be noted that even spot welds

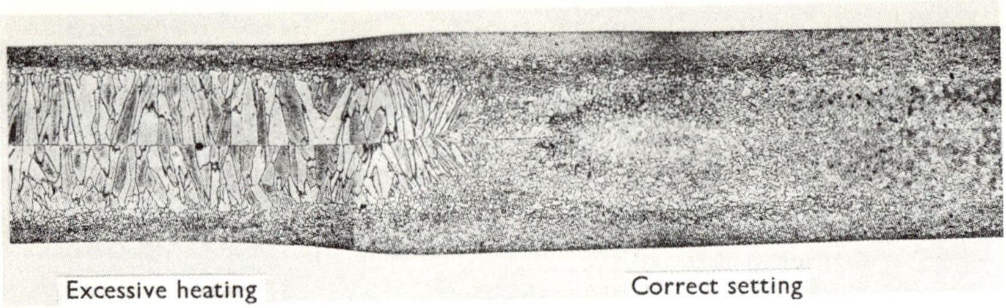

Fig. 5.10. Spot welds in 17 % chromium steel sheets (× 17).

carried out under these conditions do not show really brittle behaviour, and that it is of course impossible to perform impact tests on such welds.

The same considerations are applicable to seam welding, to the extent that this process consists of making a series of partially overlapping spot welds. A variant of this process consists of welding plates with an overlap which is less than the width of the wheels, and making the welds with a considerable pressure. This is known as 'mash welding'. The severe plastic deformation of the metal can reduce the original double sheet thickness to the thickness of a single sheet, and is a favourable factor for semi-ferritic and ferritic steels.

7.2 Butt welding of bars and sections

When this process is applied to ordinary steels, the pressure of the work-pieces against each other is generally low, and the components are brought together as a result of the expulsion of liquid metal from the joint. This method as applied to 17 % chromium steels does not give very good results, because of the final weld metal structure and because porosity often occurs in the weld. It is preferable to weld these steels with a very high pressure, so that the joint is made by forging rather than by fusion.

In any case, the upset must be sufficient to expel the regions of the metal which were originally in contact and which were oxidised during the heating cycle. Indeed, if this condition is not fulfilled, the oxide acts as a nucleus for the formation of austenite during cooling, and the resulting martensite may form a brittle plane traversing the whole joint. This is illustrated by fig. 5.11. When these precautions have been taken, the welds are quite ductile, and joints in wires can be drawn down without difficulty following heat treatment at 800 °C.

Finally, flash welding is well adapted to this type of steel, as its normal cycle involves the application of considerable pressure between the work-pieces.

7.3 Tubes rolled and welded by induction heating

The joint edges are heated by induction, and pressed together, the operation being performed continuously, and permitting high welding speeds to

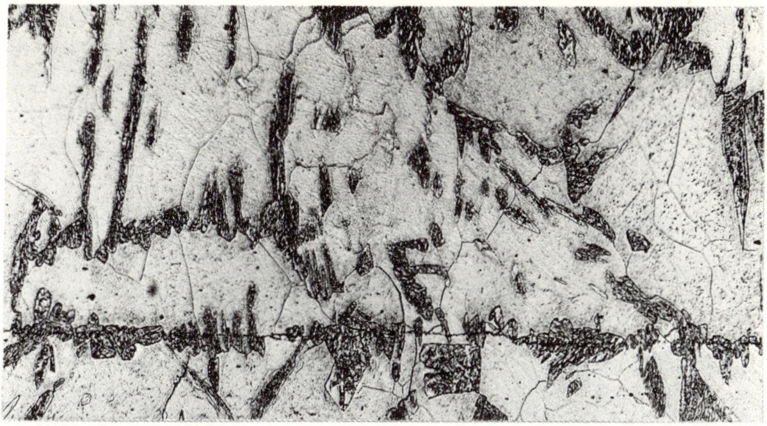

Fig. 5.11. Line of martensite resulting from oxide in a 17 % chromium steel resistance butt weld (×200).

be obtained. Heating can be very localised if the frequency is high, and since considerable deformation is applied to the solid material, the useful part of the joint contains only unmelted metal, with a fine grain size. This is illustrated in fig. 5.12, which shows a transverse section through a weld in a 0·8 mm (0·032 in.) thick tube in the as-welded condition, the external surface having been ground smooth. The upset must again be sufficient to expel the oxide formed during heating, as in §7.2.

These welds are very ductile, and the tubes can be cold-formed without heat treatment after welding.

8 Corrosion resistance of welded joints in semi-ferritic and ferritic steels

The welding thermal cycle reduces the corrosion resistance of semi-ferritic and ferritic steels. This deterioration can be revealed by the conventional immersion tests in copper sulphate–sulphuric acid, nitric acid, and nitric–hydrofluoric acid media.

8.1 Experimental observations

Semi-ferritic and ferritic steels heated to a temperature above approximately 950 °C and rapidly cooled, are sensitive to preferential corrosion at grain boundaries, in the acid media cited above. Heat treatment inducing susceptibility to this intergranular (or intercrystalline) corrosion is called a 'sensitisation' treatment, and the rapid dissolution of the metal at or immediately adjacent to the grain boundaries can lead to the dislodging of grains and swift disintegration of the metal. The problem of intergranular corrosion will be considered further in chapter 6, § 5.2, with reference to

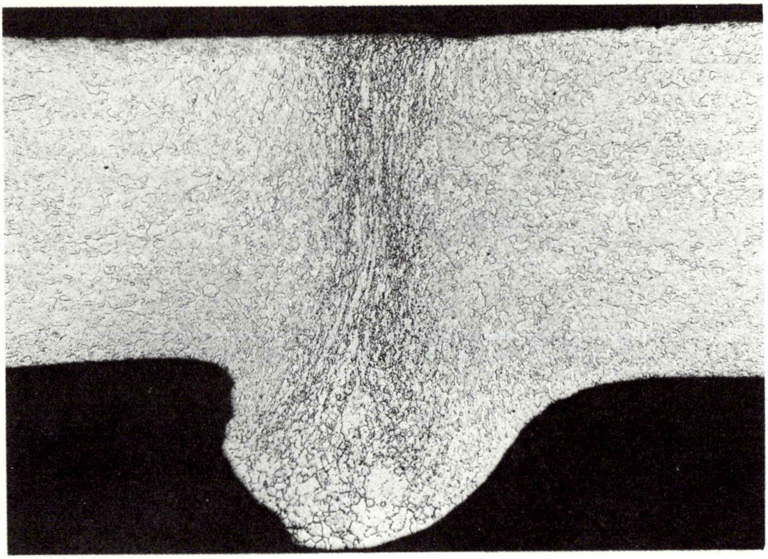

Fig. 5.12. Section of a longitudinal induction weld in a 17 % chromium steel tube (×60).

austenitic stainless steels, since it plays an important part in the welding of these materials.

In the case of welded joints in ferritic and semi-ferritic stainless steels, this corrosion is manifest in the HAZ adjacent to the weld metal as well as in the weld metal itself, although it seems less important in the latter region. The disintegration which is caused can lead to rapid collapse of the assembly.

It is possible to restore the corrosion resistance of the sensitised metal by subjecting it to a heat treatment at between approximately 650 and 900 °C. As this temperature range is below the sensitisation range, a localised heat treatment of the weld, for instance by heating with a gas torch, improves the corrosion resistance of the weld. If the sensitisation temperature were below the remedial treatment temperature, as is the case with austenitic steels, a local heat treatment would only shift the sensitised zone. Thus, heat treatment at 800 °C applied to welds in 17 % chromium steels enables the corrosion resistance of the weld to be regained.

The chromium content does not seem to have a marked effect upon this type of corrosion. A decrease in carbon content improves the intergranular corrosion resistance of joints in the as-welded condition, but sensitisation is still possible with only 0·01 % carbon. It is necessary to lower the carbon content to 0·002 % in order to suppress the tendency for sensitisation of steels containing 25 % chromium. It is also possible to reduce the sensitisation of ferritic and semi-ferritic steels by the addition of elements forming carbides or carbonitrides which are stable at high temperature [6], [7]. For example, sensitisation due to welding in a 17 % chromium steel can be avoided when the titanium content is 6–9 times its carbon content. However, this improvement does not appear in nitric acid, for which an addition of titanium to the steel has the effect only of raising the temperature above which sensitisation can take place.

8.2 Mechanism of sensitisation to intercrystalline corrosion

A number of mechanisms have been proposed. Some of them are based upon the formation of intergranular austenite at elevated temperature which is impoverished in chromium and enriched in carbon, as is the austenite formed under equilibrium conditions. The martensite resulting from this austenite could be sufficiently poor in chromium to reduce its corrosion resistance considerably. Alternatively, the austenite could lead to the formation of iron carbides on cooling, which were liable to dissolve in acid media [6]; these carbides would change into chromium carbides which were insoluble in acids during heat treatment at between 650 and 900 °C.

Other authors [7] have found that sensitisation can take place in material with such a composition that austenite does not form, and that the minimum sensitisation temperature does not depend upon the chromium content. Thus, they propose various mechanisms based on the presence of certain types of carbides or carbonitrides precipitated at the grain boundaries on rapid cooling from high temperature. The corrosion could result from the local stresses caused by this precipitation, or from the dissolution of the precipitates themselves.

In conclusion, it can be pointed out that, in France and English-speaking countries, these steels are seldom employed in the fusion-welded state, or in media where there is a risk of intercrystalline corrosion. This is not the case in Germany, where research seems to be directed towards an explanation of sensitisation by carbide or carbonitride precipitation.

9 Sigma phase

Sigma phase is an intermetallic compound appearing in the iron–chromium system, and is therefore mentioned in this chapter, although it does not play a major role in the welding of ferritic and semi-ferritic steels. The sigma phase is hard and brittle, and its presence in welds can cause a real brittleness, which is evident even in the absence of notches and does not disappear with a moderate rise of temperature. However, as ferritic and semi-ferritic steel welds have a low toughness at room temperature, transformation to sigma phase in such welds is of minor importance, although it will be seen in chapter 6, § 8.1, that the formation of sigma type phases is of considerable significance in the welding of austenitic stainless steels.

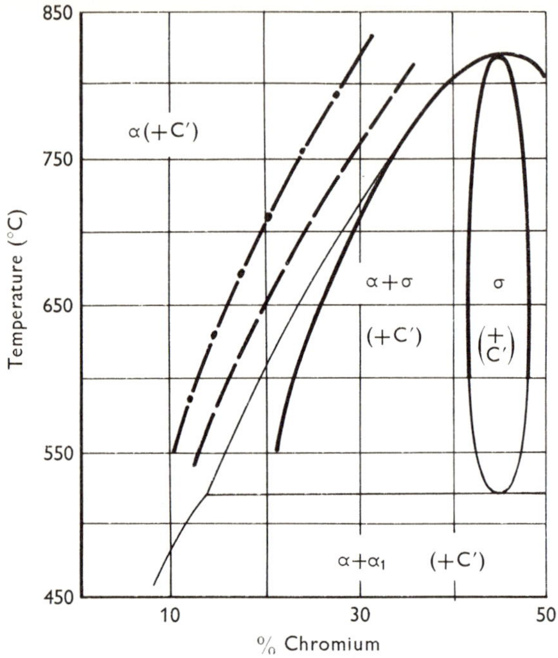

Fig. 5.13. Equilibrium diagram of the sigma phase: ═══ iron–chromium alloys; – – – steels containing up to 1 % silicon; –·–· steels containing 2·5 % silicon: α, ferrite; α_1, chromium-rich phase; σ, sigma; C′, carbide.

Sigma phase has a composition centred on a chromium content of 45 %, and can appear during prolonged heat treatment at above 550 °C of iron–chromium type alloys containing more than about 22 % chromium. The classical equilibrium diagram for sigma phase is shown by the thick lines in fig. 5.13. In fact, sigma phase can probably be formed below 550 °C, and a more complete diagram, represented by the thin lines in fig. 5.13, has been proposed [8] in which sigma phase would exist down to 520 °C; below this temperature, there would exist a range in which alpha ferrite co-exists with a phase rich in chromium [9].

It must be noted that sigma phase formation in industrial alloys requires the diffusion of chromium, and that this diffusion becomes slow when the temperature decreases. Consequently, very long heat treatments are required to study the initiation and growth of this phase at below 550 °C. The sigma range diminishes as the temperature rises, and disappears around 820 °C; with chromium contents below 45 %, the sigma dissolution temperature increases with the chromium content. Some carbides can exist in the presence of sigma phase.

Alloying and other elements present in chromium steels can displace the range over which sigma phase is stable. This is particularly the case with silicon, which is often added to ferritic steels to improve their resistance to oxidation at high temperature. Silicon considerably extends the range of existence of sigma towards higher temperatures and lower chromium contents, as is indicated in fig. 5.13[10].

References

[1] R. Tricot and R. Castro. Mémoires scientifiques, *Rev. Mét.* **63** (1966) 657.
[2] R. Tricot and R. Castro. Mémoires scientifiques, *Rev. Mét.* **59** (1962) 587.
[3] Y. Nishio, T. Ohmae, Y. Yoshida and Y. Miura. *Weld. J. Res. Sup.* **50** (Jan. 1971) 9 s.
[4] J. Lefèvre, R. Tricot and A. Gueussier. *Rev. Mét.* **68**, n°2, (1971) 95–106.
[5] R. N. Wright. *Weld. J. Res. Sup.* **50** (Oct. 1970) 434 s.
[6] E. Houdremont and W. Tofaute. *Stahl Eisen*, **72** (1952) 539.
[7] R. A. Lula, A. J. Lena and G. C. Kiefer. *Transactions A.S.M.* **46** (1954) 197.
[8] E. O. Williams and H. W. Paxton. *J. Iron Steel Inst.* **185** (1957) 358.
[9] J. Hochmann. *Rev. Mét.* **48** (1951) 734.
[10] H. J. Schuller. *Archiv Eisenhuttenwesen*, **36** (1965) 513.

6

THE WELDING OF AUSTENITIC OR AUSTENITIC-FERRITIC STAINLESS AND HEAT-RESISTING STEELS

The properties of martensitic and ferritic stainless steels resemble those of ordinary or low alloy steels. Austenitic stainless steels, on the other hand, have a number of distinctive characteristics which clearly distinguish them from all other categories of steels. Because of this, it is useful to recall certain basic concepts regarding the constitution, structure and application of these alloys, as these concepts are indispensable for the comprehension of phenomena which take place during welding operations.

It was pointed out in chapter 1 that an austenitic structure and the presence of nickel improve the corrosion resistance of chromium steels, particularly in slightly oxidising or reducing media. In addition, austenitic stainless steels are characterised by the following properties:

- Their ductility and toughness are distinctly higher than those of other steels, and these properties are retained at very low temperature.
- In general they are considerably hardened by plastic deformation; this hardening by cold-working, which occurs without causing brittleness, gives them a high mechanical adaptability.
- In the absence of cold-working, they have no elastic limit in the conventional sense of the word, and they are slightly deformed by creep under stress at room temperature.
- Their composition and structure give them good hot strength, which permits their application at temperatures which are considerably above the maximum service temperature of low alloy steels or martensitic and ferritic stainless steels.
- Their ready weldability is almost unique in the field of special steels. Indeed, it will be seen below that in most of their numerous applications, it is easy to make sound welds without preheating, while the welds can be put into service without heat treatment after welding, and have properties closely comparable to those of the base metal. However, this result can be achieved only by suitable choice of the welding process and filler metal, which must be made in accordance with the welding metallurgy of this type of steel, from the point of view of the weldability and the working

conditions of the assemblies. In this chapter, these two aspects will be considered successively.

1 General comments – composition and structure

In chapter 1, §3, it was pointed out that austenitic stainless and heat-resisting steels are alloys derived from the iron–chromium–nickel system. Their chromium content is generally above 16 %, and they contain a sufficient quantity of nickel for the austenitic structure to be stable at room temperature, or to be retained by fast cooling from elevated temperatures.

There are numerous types of austenitic stainless steel, but the most widely used alloys are steels containing 18 % chromium, 10 % nickel and at most 0·1 % carbon. This will be used as an example to describe briefly the metallography of these steels, and will then be compared with other types of composition.

1.1 18 % Cr/10 % Ni austenitic stainless steels

In this group of steels the chromium content is always about 18 %, whereas the nickel proportion can vary over a fairly wide range. Fig. 6.1 represents a schematic section of a constitutional diagram for iron–chromium–nickel alloys containing 18 % chromium and less than 0·03 % carbon. The limits of this diagram, although having been the subject of numerous studies, remain rather imprecise, and the nickel contents cited below are approximate.

Up to 10–11 % nickel, the alloy solidifies as delta ferrite, or at least with a very high proportion of this phase. Beyond this percentage, a peritectic reaction occurs, and solidification takes place to delta ferrite and austenite. It must be noted that fig. 6.1 is not a true equilibrium diagram, because the points representing compositions of the phases in equilibrium are not connected by straight lines situated in the plane of the figure. Thus, it is difficult to know accurately the nickel content above which solidification of austenite alone takes place, but it is probable that it is of the order of 14–15 %.

As long as the nickel content remains below 1–1·5 %, the alloy behaves like a fully ferritic steel during cooling from high temperatures. Above this percentage, there appears a temperature range with two phases, austenite and delta ferrite, which extends as the nickel content increases; the alloy then behaves as a semi-ferritic steel. Finally, above 3–3·5 % nickel, there is a temperature range in which the alloy is purely austenitic, and which widens as the nickel content becomes higher. With increasing nickel, the M_s temperature is lowered, but as long as the nickel content does not exceed about 7–8 %, the M_s of the austenite remains above room temperature; the steel is then of a martensitic type.

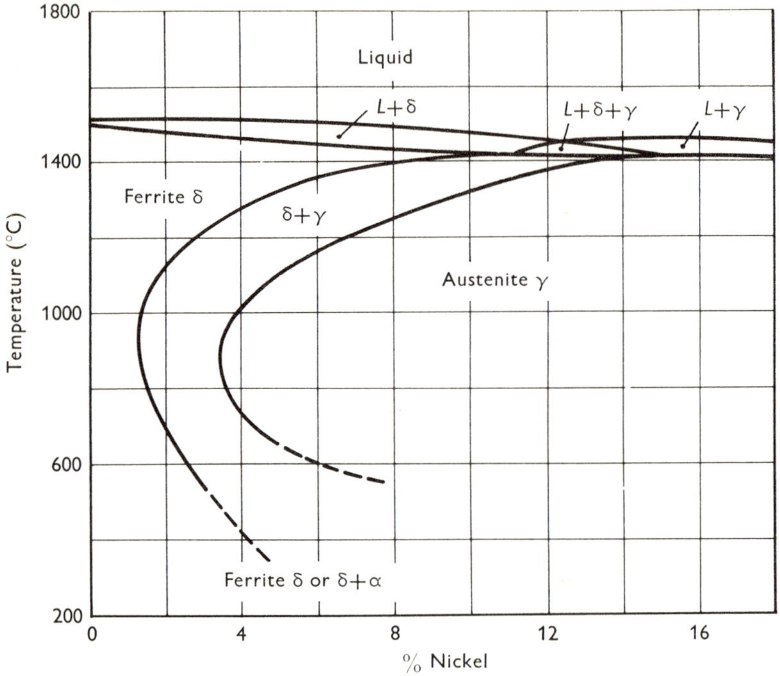

Fig. 6.1. Vertical section of very low carbon Fe–Cr–Ni alloys at 18 % chromium.

In general, austenitic steels are alloys having sufficient nickel to be fully austenitic below approximately 1150 °C, and to remain so after cooling to room temperature. In the case of alloys containing 18 % chromium, a nickel content of 8 % is in principle sufficient to fulfill these conditions. However, commercial 18 % chromium steels generally contain at least 9 % nickel for the following reasons:

– Even with 8 % nickel, the austenite can be at least partly changed into martensite by plastic deformation at around room temperature. An additional increase in nickel content improves the austenite stability and allows this transformation to be avoided, since it can sometimes cause difficulties during fabrication or service of the metal.

– Because of the relatively high mechanical strength of austenite at elevated temperature, the hot-working (rolling and forging) of these steels must be carried out at very high temperature. However, the presence of a marked quantity of ferrite in the austenite prevents this working from being carried out satisfactorily. Fig. 6.1 shows that an increase in the nickel content raises the temperature at which delta ferrite appears on heating, and reduces the proportion of this phase at a specific temperature. Thus, nickel has a favourable effect on the hot-working of the metal.

– Finally, it must be noted that, as peritectic reactions are generally slow, the structures corresponding to the physico-chemical equilibrium are rarely obtained in industrial fabrications. This results in chemical heterogeneities which can cause the formation of stable delta ferrite and unstable austenite during cooling, even though the nickel content may be theoretically sufficient to avoid this effect under equilibrium conditions. An increase in the nickel content reduces the possibility of such phenomena occurring.

A change in the chromium content has an effect similar to those which have already been encountered in other types of stainless steels. An increase in the proportion of this element extends the temperature range over which delta ferrite exists, and consequently requires an increase in nickel content to obtain a fully austenitic structure below a specific temperature. However, an increase in the chromium content also has the effect of increasing the austenite stability with regard to cooling or plastic deformation. This effect is not peculiar to chromium; as is the case with martensitic steels, an increase in the content of all alloying elements lowers the M_s temperature.

The effect of carbon upon the structure of austenitic stainless steels at above approximately 900 °C is similar to that of nickel: like nickel, it extends the austenite range, and consequently reduces the quantity of delta ferrite present at elevated temperature. However the solubility of carbon in austenite decreases as the temperature falls. If the content is below 0·03 %, this effect has little importance metallurgically. With carbon contents between 0·03 and 0·15 %, a level rarely exceeded in conventional 18 % Cr/10 % Ni alloys, carbon is entirely soluble in the steel at temperatures above 1000 °C, but during slow cooling, or even a brief stay between about 500 and 900 °C, chromium carbides (possibly containing some other alloying elements) can precipitate, and impair certain service properties of these steels, particularly their plasticity and corrosion resistance.

This is one of the reasons why austenitic stainless steels are employed after rapid cooling in air or water following solution treatment at 1000–1100 °C. The soaking at elevated temperature permits recrystallisation of the metal after plastic working operations, while the carbon goes into solid solution. The fast cooling prevents the precipitation of carbides, and gives the metal a fully austenitic equiaxed structure, with associated optimum properties of ductility and corrosion resistance.

Finally, austenitic stainless steels have the same minor elements as those encountered in other types of steels. They can also contain certain alloying elements which will be considered in §2.2, dealing with the effect of the chemical composition of the metal upon the structure of welded joints.

1.2 Other types of austenitic stainless steels

The most common of these other types of stainless steels are as follows, listed in order of decreasing tendency to form delta ferrite at elevated temperatures:

– Steels containing 17 % chromium and 7 % nickel. The austenite of these steels is relatively unstable, and this is turned to account either to facilitate deep-drawing when the carbon content is low, or to obtain very high mechanical strength by cold-working when the carbon content exceeds 0·1 %. They are rarely employed under severe corrosion conditions.

– Refractory steels containing 24 % chromium and 12 % nickel. Their chromium content provides them with a good oxidation resistance at elevated temperature, and since they are largely rendered austenitic by a carbon content above 0·15 %, they avoid the disadvantages encountered with 25 % chromium ferritic steels. They often contain more than 1 % silicon to improve their high temperature corrosion resistance.

– Refractory steels containing 25 % chromium and 20 % nickel. These alloys have an even better oxidation resistance at high temperature. They are fully austenitic regardless of their carbon content, and, like the preceding alloys, can have high silicon contents.

– Refractory steels containing 18 % chromium and 38 % nickel. These are fully austenitic at all temperatures, and their corrosion resistance at elevated temperature covers a large range of applications. Like the previous alloys, they can contain more than 1 % silicon; their carbon content varies between 0·05 and 0·15 %.

– Finally, a category of stainless steels exists in which an austenitic structure is obtained by the presence of sufficient manganese (generally exceeding 5 %), to increase the stability of the austenite on cooling. These steels must contain a little nickel, or nickel and nitrogen, to possess a fully austenitic structure at high temperature.

2 Microstructural changes in welding in 18 % Cr/10 % Ni type steels

2.1 The austenite–ferrite structure

Weld metals in this type of steel generally have a mixed austenite–ferrite structure after cooling to room temperature, which is quite different from the fully austenitic structure of metal hyperquenched from 1000–1100 °C. This mixed structure is due mainly to the fast cooling of the weld metal. As shown above (§1.1), during cooling from the liquid state these steels traverse a temperature range in which delta ferrite is present in substantial proportions (which can go up to 100 %). The transformation of this ferrite into austenite during cooling is completed only if the steel remains for a sufficiently long time in the range of temperature over which rapid transformation occurs. This is generally not the case with welds, where the rapid cooling, which is effectively a quenching operation, causes the retention of ferrite down to room temperature, the amount of ferrite reflecting the much higher proportion existing at elevated temperature. This quantity of ferrite depends upon the cooling rate, and is greater as cooling is faster in the range above 1250 °C.

Further, segregation during solidification causes local compositional variations, as was recalled in chapter 2, §4.1. This chemical heterogeneity increases the ferrite stability, and retards the regression of delta ferrite into austenite during cooling in certain regions of the dendritic solidification structure. Thus, segregation governs the distribution of ferrite present in the weld metal after cooling is complete, as the ferrite in these regions is obviously the slowest to transform.

Although not an equilibrium diagram, the shape of the liquidus and solidus curves in fig. 6.1 suggests that the dendrite axis is impoverished in nickel, and consequently the ferrite in these regions is the most stable. On the other hand, the last liquid to solidify may be enriched in chromium, or in elements having an influence upon the structure similar to that of chromium (§2.2). This would tend to increase the stability of the delta ferrite corresponding to the interdendrite areas. In fact it has been possible to demonstrate that the segregation in 18 % Cr/10 % Ni alloys during solidification can cause the simultaneous formation of ferrite in the dendrite axes, and in the interdendritic intervals [1].

Figure 6.2 shows the microstructure of an austenitic–ferritic weld metal in an 18 % Cr/10 % Ni type steel, and clearly reveals the dendritic appearance of the ferrite distribution. It is sometimes difficult to determine the preferential distribution (dendritic axes or interdendritic regions) of ferrite solely from a metallographic section of a weld. The distribution depends upon the metal composition, the relative segregation rates of the various elements, and the relative transformation rates of the ferrite corresponding to different regions. Morphologically it can be pointed out that phases localised in the interdendritic regions usually have concave contours.

The distinctive properties of austenitic stainless steel welds result from this structure. As the austenite constituting the greater part of the weld metal does not transform into martensite, there is no risk of the type of cracking caused by the formation of large quantities of martensite, and joints in the as-welded condition possess excellent ductility comparable to the base material. Moreover, the austenite–ferrite aggregate does not suffer rapid grain coarsening at elevated temperature. Both of these characteristics, and particularly the first one, greatly facilitate the welding of austenitic steels, and there is no necessity to apply preheat to a weldment, nor to carry out heat treatment after welding.

In general, the carbon content of 18 % Cr/10 % Ni austenitic stainless

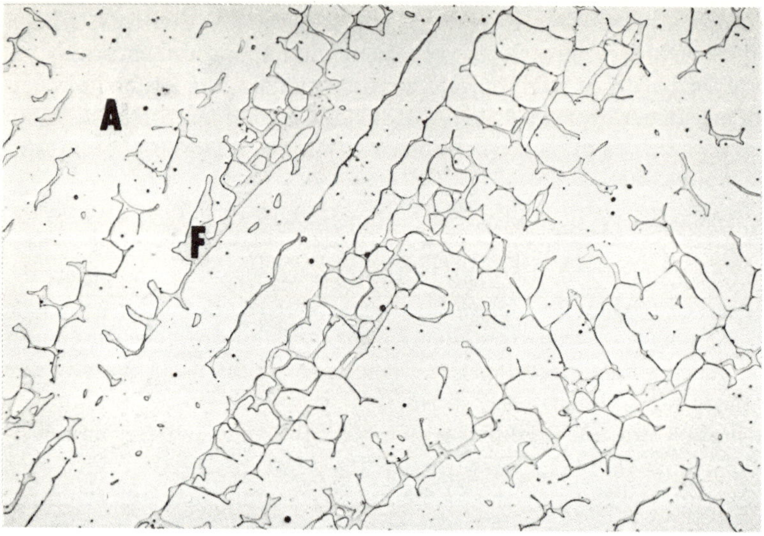

Fig. 6.2. Austenitic–ferritic weld metal in an 18 % Cr/10 % Ni type steel (× 450): A, austenite; F, ferrite.

steels is low enough for this element to be entirely soluble in the metal above 1000 °C. Welds in these steels may, however, contain some carbides, because precipitation can take place during cooling of the metal heated to a very high temperature, or during the thermal cycle experienced by certain regions of the HAZ. Such precipitation occurs at grain boundaries, and this phenomenon will be considered in detail in §5.2, dealing with the intercrystalline corrosion of welds.

2.2 Effect of the chemical composition upon the weld structure – the Schaeffler diagram

It is possible to group the minor and alloying elements present in austenitic stainless and heat-resisting steels in two categories. The elements of one group have an effect upon the structure similar to that of nickel, and widen the range of austenite existence; these elements are called 'austenite formers'. The other group contains elements which, like chromium, enlarge the delta ferrite range, and are thus termed 'ferrite formers'.

Apart from nickel and carbon as mentioned in §1.1, the group of austenite formers includes the following elements:

Nitrogen, which may be present in amounts up to 0·07 % as a minor element. Nitrogen can also be employed as an alloying element either partly to replace nickel or to increase the mechanical strength of 18 % Cr/10 % Ni steels: in the latter case, it plays a role similar to that of carbon, without the associated disadvantages related to the precipitation of carbides.

Manganese, which is generally considered an austenite former. It is, in fact, possible to retain a purely austenitic structure at room temperature by partially replacing nickel by a larger quantity of manganese [2], although it seems that at above about 4 %, the influence of manganese upon the structure at very high temperature may be comparable to that of ferrite formers [3]. As a minor element, the manganese content is generally below 2·5 %; it may be present in a larger quantity either to allow a reduction in nickel content, or to improve the weldability of certain austenitic steels (§4.4).

Copper, which seems to be a weak austenite former. This element improves the corrosion resistance in certain environments and also increases the high temperature creep resistance of 18 % Cr/10 % Ni type alloys.

The group of ferrite formers contains chromium together with the following elements:

Silicon, which as a minor element is below 1 %. It can be employed as an alloying element in refractory steels, or to improve the resistance of 18 % Cr/10 % Ni grades in certain cases of stress corrosion.

Molybdenum, which improves the corrosion resistance of austenitic steels in various humid conditions. It also increases the mechanical strength of 18 % Cr/10 % Ni steels at high temperature.

Tungsten, which improves the mechanical strength at elevated temperature.

These last two elements favour the formation of complex carbides.

Titanium and niobium, which are used to avoid the sensitisation of austenitic stainless steels to intercrystalline corrosion. These elements effectively fix the carbon by forming carbides at high temperature which are little soluble in the matrix (§5.2). They also improve the creep resistance at high temperature.

The structure of a weld after complete cooling depends upon the quantity of ferrite which existed at very high temperature, and consequently upon the relative proportions of austenite formers and ferrite formers. L. A. Schaeffler [4] studied quantitatively the effect of these elements upon the structure, and particularly upon the ferrite content of weld metals deposited under well-defined thermal conditions. The results were combined in a diagram named after him, and represented in fig. 6.3. The coordinate axes are as follows:

The nickel equivalent, which is the sum of the nickel content and the contents of other austenite formers, multiplied by coefficients representing their effects as compared to that of nickel. The strong austenite-forming tendency of carbon is expressed by a very high coefficient. It seems that manganese should be included in the nickel equivalent, with a coefficient of 0·5, only when its content does not exceed approximately 2·5 %, i.e. within the conditions for which the Schaeffler diagram was established.

The chromium equivalent is calculated in an analogous manner.

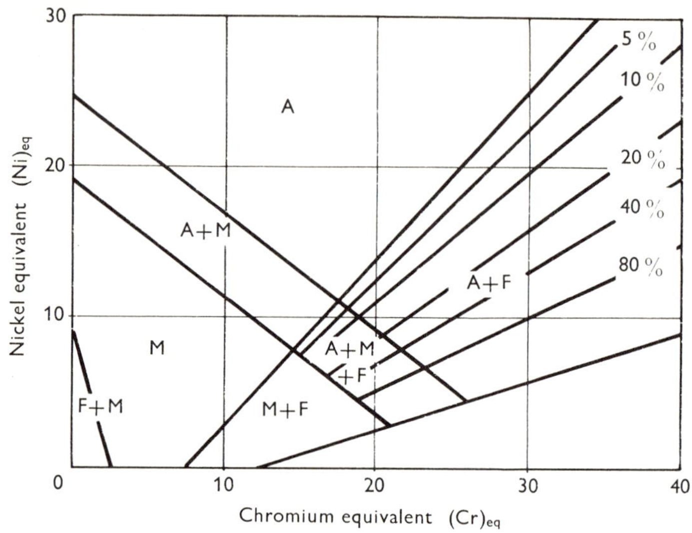

Fig. 6.3. Schaeffler diagram.
Nickel equivalent $(Ni)_{eq}$ = % Ni + 30 × % C + 0·5 × % Mn.
Chromium equivalent $(Cr)_{eq}$ = % Cr + % Mo + 1·5 × % Si + 0·5 × % Nb.
A, austenite; F, ferrite; M, martensite.

The diagram permits the prediction of the microstructure of a weld metal, when its chemical composition is known. The domain of austenitic stainless steels corresponds to a nickel equivalent above 8 and a chromium equivalent above 16. It can be seen that after cooling, welds with nickel and chromium equivalents above these values can have various structures, consisting of austenite, austenite with a little quantity of martensite, austenite and ferrite with some martensite, or austenite and ferrite. In this last range, which concerns the most common austenitic stainless steels, the curves for a constant ferrite proportion are straight lines. Thus, this diagram enables the proportion of ferrite present in a weld metal to be estimated from its chemical composition, and as shown below, this has important practical consequences.

The Schaeffler diagram is not restricted to austenitic steels, and it permits the prediction of the structures that are encountered in the stainless steel welds described in chapters 4 and 5, such as martensite with a little quantity of ferrite, ferrite with some martensite, or pure ferrite.

D. Séférian[5] has proposed a convenient formula derived from the Schaeffler diagram, which allows the calculation of the ferrite content of a weld deposit with such a composition that the structure is austenitic–ferritic:

$$\% \text{ ferrite} = 3\,[(Cr)_{eq} - 0{\cdot}93\,(Ni)_{eq} - 6{\cdot}7].$$

The Schaeffler diagram does not take account of nitrogen, which has a marked influence as an austenite former. De Long[6] has proposed a

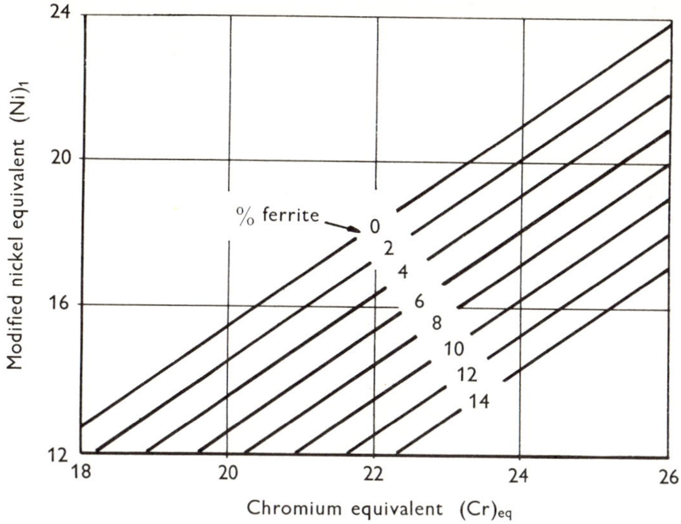

Fig. 6.4. Diagram of W. T. De Long. $(Ni)_1 = (Ni)_{eq} + 30 \times \%$ N.

modification to part of the diagram, as shown in fig. 6.4, where nitrogen is included in the calculation of the nickel equivalent with a coefficient equal to 30, identical to that of carbon. This diagram was established for molybdenum-free compositions.

Other elements can also be taken into account in the original diagram and Séférian's formula. Copper can be included in the nickel equivalent, with a coefficient of about 0·6, and nitrogen with a coefficient between 10 and 25. Allowance for tungsten can be made in the chromium equivalent, with a coefficient of about 0·5, and titanium with a coefficient between 2 and 5.

Finally, it must be noted that for a given composition, the ferrite content decreases with the cooling rate, and the information furnished by the Schaeffler diagram must not, therefore, be considered as absolute. This diagram permits the calculation of a *ferritic tendency*† which is equal to the ferrite content of weld metals deposited under the same thermal conditions as those used to establish the diagram. Strictly speaking, this ferritic tendency would allow a comparison only between the ferrite contents of welds carried out under identical conditions. However, experience has shown that the Schaeffler diagram enables the ferrite content of welds to be predicted with a sufficient precision for the majority of cases in practice.

Welds in 18 % Cr/10 % Ni type steels, or in types derived from them by an addition of other alloying elements mentioned above, generally contain

† *Note:* The authors use 'ferritic tendency' generally throughout the book. This term is not common in U.K. practice, and the more usual, although less precise, term 'ferrite content' has been adopted in the translation, when this is more appropriate.

		Fe	Ni	Mn	Cr	Si
Austenite:	dendrite axes	68·6	11·4	1·6	18	0·2
	interdendritic regions	64·9	12·8	2·1	19·4	0·5
Ferrite:	dendrite axes	66·5	5·5	1·4	26	0·45
	interdendritic regions	63·2	6·2	1·7	28·1	0·6

between 0 and 20 % delta ferrite after final cooling. It will be seen below that it is often necessary to maintain the ferrite proportion between quite precise limits, for reasons related to the weldability of this type of steel, or to the service conditions of welded joints.

2.3 Chemical composition of austenite and ferrite

Austenite and ferrite do not have the same composition, and in general ferrite contains more ferrite formers whereas austenite has more austenite formers. This can be shown by electron probe microanalysis techniques. As an example, the table above gives the ferrite and austenite compositions present in the dendrite axes and interdendritic regions of a forged stainless steel containing 18·6 % chromium, 11·8 % nickel, 0·12 % carbon, 0·42 % silicon, 1·72 % manganese and 0·49 % titanium.

The results clearly illustrate the tendency for ferrite to be enriched in the ferrite formers chromium and silicon, and impoverished in nickel and manganese. They also show that the phases present in the interdendritic regions contain more alloying elements than the same phases present in the dendrite axes, as a result of the segregation on solidification considered in §2.1.

It is more difficult to obtain quantitative analyses of weld metals because the rapid solidification generally gives rise to very fine two-phase microstructures. However, it is possible to obtain results giving an idea of the distribution of alloying elements in the structure, as shown in fig. 6.5. The micrograph is of a weld metal in steel containing 18 % chromium and 9·9 % nickel. A colouring etchant reveals the dendritic structure and small pools of ferrite. The darker track between the two black points is the trajectory of the electron probe, and above the micrograph the chromium and nickel contents measured along this trajectory are shown. It is apparent that, in this weld, the interdendritic segregation of nickel is well pronounced, while that of chromium is limited. It will also be noted that, when the probe passes over the ferrite island situated towards the right-hand side of the figure, the nickel content falls to 7 %, while the chromium content exceeds 20·5 %. These contents are in fact too high for nickel and too low for chromium, because the area examined by the probe, indicated by the size of the black points, is considerably larger than the ferrite particle.

Tests carried out in the same way have shown that in a weld metal

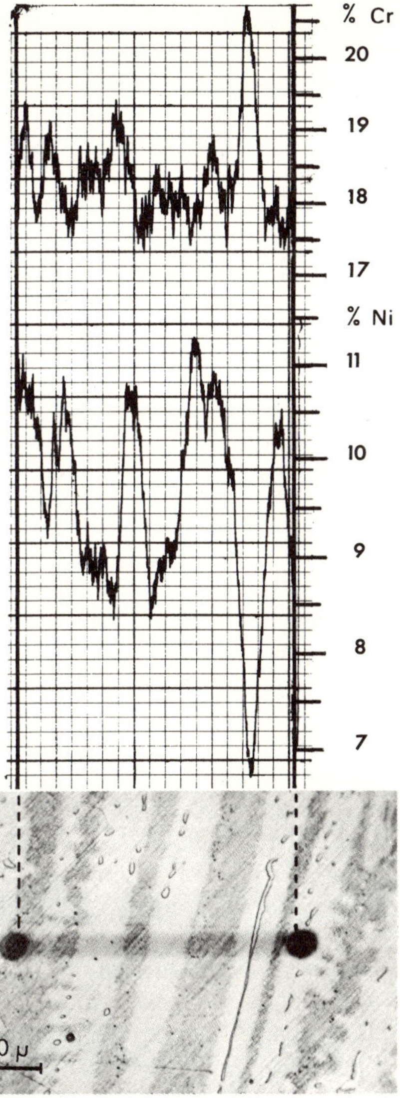

Fig. 6.5. Variations in nickel and chromium contents in the weld metal of a weld in a steel containing 18 % chromium and 9·9 % nickel.

having an average composition of 16·5 % chromium, 2 % molybdenum and 9 % nickel, the ferrite contains at least 19·5 % chromium and 4 % molybdenum, and at most 5·5 % nickel.

These differences in composition may have practical consequences in certain cases of corrosion and elevated temperature application.

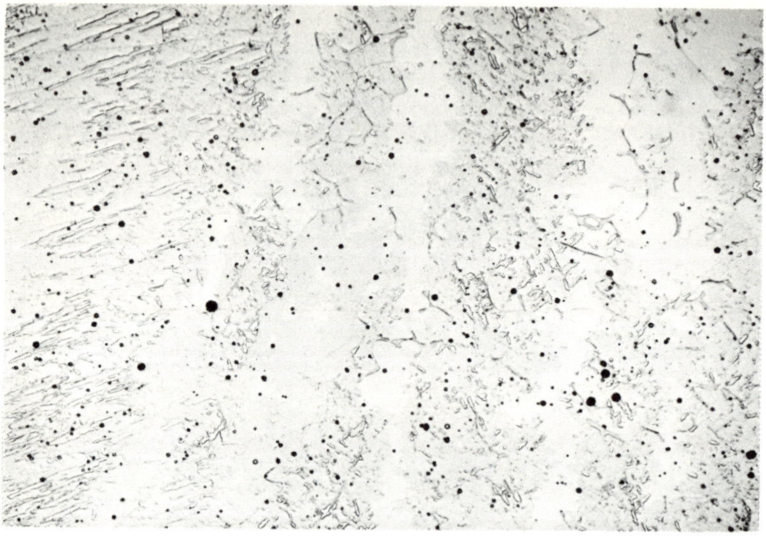

Fig. 6.6. Austenite–ferrite–martensite structure of a weld in a steel containing 17·5 % chromium and 7·5 % nickel (×450).

3 Microstructure of welds in other types of austenitic stainless and heat-resisting steels

The Schaeffler diagram permits prediction of the final structure of weld metals of the alloys enumerated in §1.2.

3.1 Austenitic steels less alloyed than the 18 % Cr/10 % Ni type

The Schaeffler diagram indicates that a simultaneous reduction of nickel and chromium equivalents in the ratio given by the Séférian formula allows retention of an austenite–ferrite structure without changing the ferritic tendency. However, the reduction in the content of alloying elements can, as a result of segregation during solidification, favour the formation of islands of unstable austenite, which may transform into martensite on cooling. This can occur during autogenous welding of 17 % Cr/7 % Ni type steels, or when service conditions necessitate the use of a filler metal containing a low proportion of ferrite formers (see §8.2). In all cases, the extent of transformation is very limited, thus avoiding any risk of cold cracking and permitting the excellent ductility characteristic of austenitic stainless steels to be retained. Fig. 6.6 is an example of such a structure.

3.2 Refractory 24 % Cr/12 % Ni type steels

The structure of weld metals in 24 % Cr/12 % Ni type steels is similar to that of welds in 18 % Cr/10 % Ni alloys, because their carbon content is often quite high, resulting in a ferrite content generally between 0 and 10 %. This carbon content also favours precipitation of carbides at grain boundaries in the weld metal and HAZ.

3.3 Refractory 25 % Cr/20 % Ni type steels

Welds in steels containing 25 % chromium and 20 % nickel are generally entirely austenitic at room temperature. However, it is possible for certain alloys of this type with a high relative proportion of ferrite formers to pass through a range in which a little ferrite is formed during cooling. This is especially the case with high silicon grades of steel. As a result, very small amounts of this phase may be present after cooling, apparently localised in the interdendritic regions. Fig. 6.7 shows the fully austenitic structure of a 25 % Cr/20 % Ni type weld metal, where the grain boundaries of the structure are 'punctuated' by carbides.

3.4 Refractory 18 % Cr/38 % Ni type steels

Welds in refractory steels containing 18 % chromium and 38 % nickel are entirely austenitic at all temperatures. Carbide precipitation at the grain boundaries is generally more pronounced than in 24 % Cr/12 % Ni and 25 % Cr/20 % Ni alloys, because the solubility of carbon is reduced as the nickel content is increased.

4 Practical weldability of austenitic stainless and heat-resisting steels

4.1 Cracking at high temperature

Because of their structure, welds in austenitic steels are of course insensitive to the type of cracking encountered in martensitic steels. However, in certain cases, austenitic welds are susceptible to a particular kind of cracking, having the two following distinctive characteristics:

- Cracking takes place at very high temperatures during cooling, probably above 1200 °C. In this range of temperature, the metal structure is not well known, and this complicates the study of such cracking.
- Cracks are interdendritic, i.e. they follow the boundaries of the solidification structure.

Figure 6.8, is an example of this type of cracking in an austenitic steel containing 17 % chromium and 12 % nickel. Cracking can affect a large part of the weld metal, and seriously impair the service properties of the assembly, as indicated by the example given in fig. 6.9, showing a section through a cracked weld in a 25 % Cr/20 % Ni steel.

It is known that fully austenitic weld metals are generally the most sensitive to cracking. All other conditions remaining the same, susceptibility to cracking at high temperature decreases as the ferritic tendency of

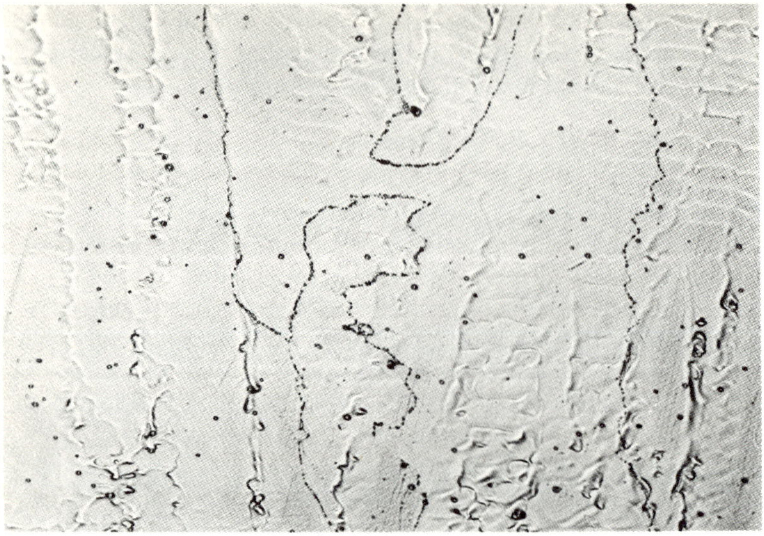

Fig. 6.7. Austenitic structure with a network of carbides in a weld in a 25 % Cr/20 % Ni/0·1 % C steel (×450).

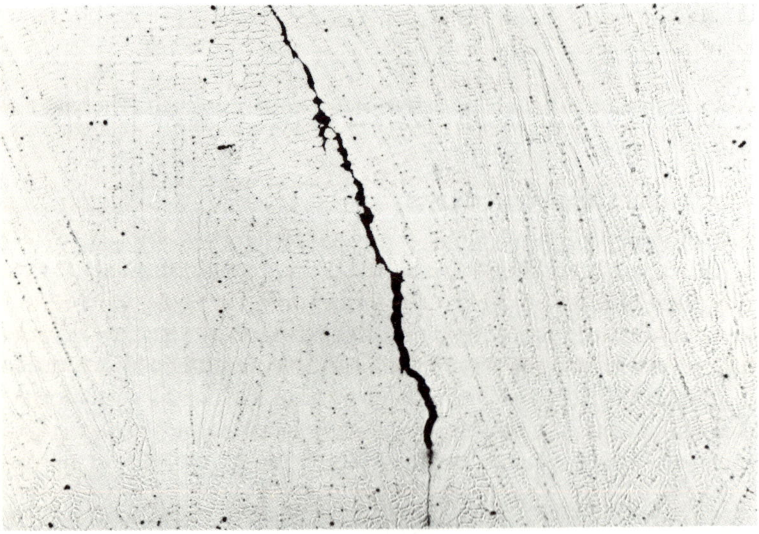

Fig. 6.8. Interdendritic crack in a fully austenitic weld in a steel containing 17 % chromium and 12 % nickel (×90).

the weld metal increases, and it has been shown experimentally that, under specific conditions of restraint, there is a critical proportion of ferrite remaining in the weld metal after cooling, above which the metal becomes insensitive to cracking.

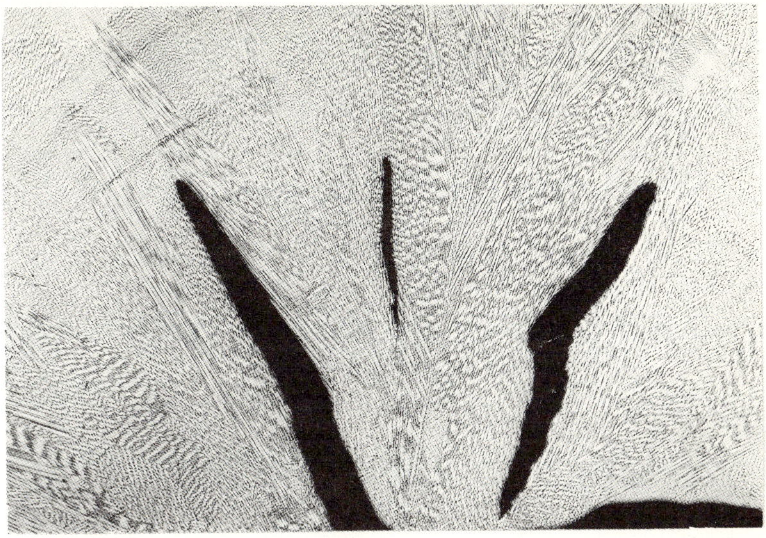

Fig. 6.9. Section through a cracked fully austenitic weld in a steel containing 25 % chromium and 20 % nickel (×20).

The quantity of ferrite necessary to obtain this immunity depends upon the joint restraint, and upon the nature and quantity of alloying elements, additional elements and impurities present in the metal. Under normal welding conditions for 18 % Cr/10 % Ni (or less alloyed) and 24 % Cr/12 % Ni types of steel, this minimum ferrite content is generally between about 1 and 8 %.

Because of their composition, 25 % Cr/20 % Ni and 18 % Cr/38 % Ni type steels are entirely austenitic during welding, and are quite susceptible to cracking.

The zone affected by the welding heat can also be sensitive to cracking at elevated temperature. In the immediate vicinity of the fusion boundary, intergranular cracks form at grain boundaries perpendicular to the direction of the shrinkage stresses. This is illustrated by fig. 6.10. It is also possible to eliminate this cracking by inducing a sufficient ferritic tendency into the base metal.

4.2 Metallurgical causes of weld metal cracking

The fundamental cause of cracking obviously lies in the existence of shrinkage stresses, but under similar conditions of shrinkage stresses, metallurgical variables may render certain stainless steels more sensitive to cracking than others.

In view of the distinctive characteristics of this kind of cracking, its causes have been sought in the physico-chemical phenomena taking place

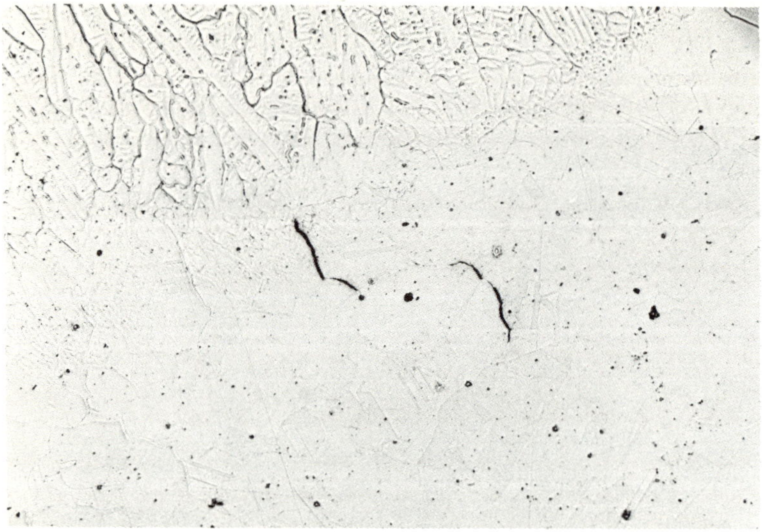

Fig. 6.10. Intergranular cracks in the heat-affected zone of a weld in a steel containing 18 % chromium and 15 % nickel (× 180).

at high temperature, and in fact, all proposed mechanisms accept that cracking at elevated temperature is due to segregation during solidification. This has the effect of the remaining liquid being enriched in certain elements as solidification progresses, and as the shrinkage stresses increase.

Among the alloying and minor elements, experience and experimental studies [7, 8] have shown that:

– Nickel, tantalum, silicon, titanium and niobium, in order of increasing detrimental effect, increase susceptibility to cracking. It is, however, preferable that silicon is maintained between certain limits. When the silicon content is below 0·3 %, the metal is too viscous and the weld bead may contain non-metallic inclusions which have not been floated out. If the silicon level exceeds 0·7 %, the risk of hot cracking increases, and the high fluidity of the molten metal can cause the weld pool to spread laterally with reduced penetration normal to the material surface, especially with TIG welding [9].

– Copper is frequently cited as causing cracking, particularly when local enrichment occurs, such as that caused by friction of plates on copper guides. The presence of copper at the root of cracked zones has sometimes been observed by microprobe analysis. Some authors [10] think that its detrimental effect is increased when the manganese content is high.

– Carbon has a strong effect in fully austenitic steels [11], the cracking sensitivity being low for $C \leqslant 0{\cdot}02$ %, and increasing with the carbon content to a maximum at about $C \simeq 0{\cdot}08$ %. For high carbon contents

(C > 0·2 %), the danger of cracking disappears, as a carbide eutectic is formed in the final stages of solidification, in sufficient amounts to give a 'self healing' effect (see § 4.5 of this chapter, and chapter 7, § 4.3).

– Chromium, tungsten, manganese and nitrogen reduce the cracking tendency.

The role of nitrogen has been particularly studied since small amounts (0·1–0·2 %) of this element are added to certain austenitic steels, in order to improve their mechanical properties. It has been found that for an equal weld metal ferrite content (measured when cold) an austenitic steel with nitrogen is less sensitive to cracking than a conventional alloy [12]. This may be due to the fact that the austenite-forming effect of nitrogen would be reduced at high temperature, hence causing a higher ferrite content when the material is still hot, with an associated lower risk of cracking.

The beneficial role of manganese may also be explained in a similar way. At low or moderate levels, this element could act as an austenite former at high temperature, this action being reversed for contents above 6–8 %. On the other hand, some authors [13] put forward the property of this element of modifying the relationship between the liquid–solid and solid–solid interfacial energies. In this model, wetting of the material by the last traces of liquid does not occur: the liquid film breaks prematurely, and the residual liquid remains at triple points in the structure, thus reducing the sensitivity to hot cracking.

In molybdenum steels for corrosion resistant duty, the austenitising effect of manganese on the structure after solidification leads to the disappearance of high temperature ferrite during cooling to room temperature. The resultant fully austenitic material does not suffer the selective corrosion of retained ferrite, which is observed at low manganese contents [14].

Among the additional and impurity elements, zirconium, boron, sulphur and phosphorus increase the cracking tendency, while oxygen would also be included in this category.

Many explanations have been proposed to explain the detrimental role played by segregation of the various elements.

It has been suggested [10], [15], [16] that the segregation of certain elements causes the formation, at the end of solidification, of a low melting point interdendritic liquid film. This results in a lack of cohesion, at the same time as shrinkage stresses make themselves felt. On this basis it is possible to explain the adverse effects of phosphorus, boron and, especially, sulphur which forms low melting point eutectics and sulphides, particularly in the presence of nickel. This is also the case for niobium, which forms complex eutectics with carbon, nitrogen, iron, silicon and, perhaps, manganese.

The deleterious action of silicon upon the cracking tendency of chromium–nickel austenitic steels can also be explained in this way, because the enrichment of the interdendritic liquid in silicon, nickel and manganese, can appreciably lower its solidification temperature. This is illustrated by studies carried out with the electron probe microanalyser on a highly crack-sensitive weld metal containing 19 % chromium, 15·5 % nickel, 1·8 % manganese and 3·8 % silicon. It was found that, in general, the metal corresponding to the interdendritic regions was enriched in silicon, manganese and nickel, and that the cracking seemed to be related to the formation of an intermetallic compound containing 10 % silicon, 10 % manganese, 28 % nickel, 17 % chromium and 35 % iron.

As a general rule, the tendency to form a continuous, or at least widely spread, interdendritic liquid film, while only a small quantity of liquid metal is left, depends upon the solid–liquid surface tension and the interfacial energy of the solidified metal boundaries. These energies obviously depend on the composition of the base metal, but they are no doubt also greatly influenced by the presence of additional minor elements, and by the partition of impurities between the solid and liquid metal.

Other explanations invoke a brittleness of the solidification structure in the solid state. Cracking always arises from segregation, but segregation is considered to have the effect only of reducing the mechanical strength of the grain boundaries at high temperature as compared to that of the grain centres, by for instance, favouring the separation of compounds such as carbides, silicates, oxides, or intermetallic phases.

From this point of view, it has been found [8] that nickel and niobium have an unfavourable effect because they strengthen the matrix at very high temperature, whereas chromium, titanium, manganese, molybdenum and tungsten have the opposite effect. It has also been noted that the presence of deoxidising elements promotes resistance to cracking in fully austenitic welds [17], which suggests that oxygen plays an important part in the formation of fusible or brittle interdendritic films.

These two explanations are in fact complementary, and it is very difficult when metallographic examination of a crack-sensitive weld reveals the presence of an interdendritic film, to know whether the cracking results from the low solidification temperature of this film, or from the brittleness at elevated temperature which it causes once solidification is complete. Moreover, both these effects can be additive for certain elements. This is probably the case with carbon, which renders certain refractory austenitic steels containing up to 0·3 % of this element very sensitive to cracking. In this case, the enrichment of the interdendritic liquid in carbon and the concomitant lowering of its solidification temperature are certainly factors promoting cracking, but the network of carbides that covers the boundaries of the solidification structure no doubt also plays an important role by reducing the capacity for plastic deformation at high temperature.

4.3 Metallurgical causes of cracking in the heat-affected zone

Figure 6.10 is an example of HAZ cracking. Similar cracks can also arise in the weld metal during the deposition of a subsequent run, and can then extend in this new run. The explanations given for this type of cracking are again based on the enrichment of the grain boundaries of the underlying structure in alloying elements or impurities, causing a local reduction in solidification temperature, or an embrittlement of the structure.

If the underlying metal is also a weld deposit, this segregation results from solidification; if the underlying metal is a forged metal (i.e. an HAZ), it can arise from the fact that the grain boundaries are generally preferred sites for impurities. The welding operation can accentuate this segregation [18], since heating to very high temperature leads either to recrystallisation or to grain growth of the metal. Both these phenomena involve migration of grain boundaries, and during this migration, the grain boundaries 'sweep up' impurities present in the metal en route. It is possible that this may cause very high local concentrations of impurities.

Alternatively, since the commencement of fusion gives rise to a solute-rich liquid phase usually at the grain boundaries of the structure, it may be postulated that fusion and local liquation are preceded by a migration of solute atoms towards the grain boundaries.

These phenomena have sometimes been called 'dehomogenisation'.

Finally, the grain boundaries of the HAZ can be 'decorated' with embrittling precipitates, such as carbides, nitrides, oxides, silicates or intermetallic phases, which render them incapable of supporting the stresses due to welding. Moreover, the welding thermal cycle can be successively the cause of precipitation, then of fracture, particularly when the steel is likely to precipitate carbides or intermetallic compounds.

4.4 Metallurgical remedies

Remedial measures based on metallurgical considerations are intended to improve the cracking resistance of welds under specific conditions of restraint. In view of the dominant role of restraint, technological precautions taken to reduce its severity may be particularly effective. However, it should be noted in this connection that cracking in austenitic steel welds takes place at high temperature. Thus, preheating is practically ineffective, because the normal preheating temperatures can only slightly modify the cooling rate and the intensity of shrinkage stresses at the temperature where cracking takes place.

The most commonly employed remedy with 18 % Cr/10 % Ni and 24 % Cr/12 % Ni steels stems from the experimental fact mentioned in § 4.1, and consists of adjusting the weld metal composition so that it contains between 1 and 8 % ferrite as necessary. This ferrite content is, in any event, only the result of the much higher proportion existing at the temperature range over which cracking takes place. The beneficial role of ferrite can be explained in several ways.

In general, the additional elements and impurities which have an adverse effect upon the cracking resistance are more soluble in ferrite than in austenite. When the weld metal composition is such that a considerable proportion of ferrite is present in the interdendritic regions where impurities have concentrated by segregation, the deleterious elements dissolve in the ferrite. The formation of low melting point or embrittling films is thereby avoided. This ferrite distribution is mainly encountered in compositions which are relatively rich in alloying elements.

However, ferrite can exert a favourable influence even when it remains only in the dendrite axes after cooling; this is particularly the case for welds in 18 % Cr/10 % Ni austenitic steels with a ferrite content not exceeding 4–5 %. This can be explained on a similar basis to that previously, by assuming that the ferrite present in the interdendritic spaces at very high temperature has totally transformed during cooling. It is also possible that the impoverishment in nickel (and perhaps the enrichment in chromium in some instances) of the first crystals to solidify favours the dissolution of elements which segregate interdendritically and are likely to facilitate cracking. A mechanism of this type has been suggested as accounting for the favourable influence of a slight ferritic tendency in 18 % Cr/10 % Ni steels upon the cracking due to oxide and especially silica films [11]. In an 18 % Cr/10 % Ni alloy, the first solidified crystals would be chromium-rich, permitting them to dissolve more oxygen; on the other hand, in fully austenitic material with 16 % chromium and 13 % nickel, segregation would take place in the opposite direction, facilitating the concentration of oxygen and silicon in the interdendritic regions, and thus allowing the formation of the films responsible for cracking.

The advantages of an austenite–ferrite structure from the cracking point of view can also arise from physical causes. For example, at high temperature the resistance to deformation of ferrite is much less than that of austenite: if the ferrite is present in sufficient quantity, the shrinkage stresses are relieved more rapidly.

Moreover, solidification of a fully austenitic composition gives a strongly columnar, coarse structure, in which the boundaries have a relatively small surface area. Experience has shown that austenite–ferrite structures are generally finer, while the presence of a second phase further increases the surface area of the boundaries. The result is that, all other variables remaining constant, the risk of forming fusible or brittle continuous films is reduced. This effect, however, is not necessarily specific to ferrite.

The amount of ferrite necessary to avoid cracking depends upon the weld metal composition. In an 18 % Cr/10 % Ni alloy with no deliberate addition of crack-promoting elements, cracking is generally avoided when the ferrite content exceeds 2–3 %, provided that the material is sufficiently deoxidised, and contains less than 0·6 % silicon, and 0·03 % sulphur and phosphorus. This ferrite content can be reduced when the weld metal contains 1–3 % tungsten and especially molybdenum. With higher silicon, sulphur and phosphorus contents, or when the metal contains other detrimental elements, the ferrite content must be raised in order to avoid cracking. For instance, in austenitic stainless steels containing 0·6–1 % niobium, it must often exceed 5 % for sound welds to be obtained. Fig. 6.11 is an example of such a weld, in which the presence of about 6 % ferrite has not prevented cracking.

This necessity to increase the ferrite content can be explained either by the fact that the solubility of the unfavourable elements in the ferrite is limited, or that a very high concentration of these elements in interdendritic ferritic islands can lower the solidification temperature of this phase sufficiently to render it ineffective.

There exist situations which will be referred to in the following paragraphs, in which the quantity of ferrite allowable in 18 % Cr/10 % Ni steel welds is restricted for reasons associated with the magnetism of welds, or their mechanical properties, or their resistance to corrosion. It is usually possible to find a compromise between these requirements and the necessity of avoiding cracking when the weld metal does not contain deleterious elements, because in this case, ferrite contents of about 2 % are sufficient, particularly when the weld metal contains 2–3 % molybdenum. Such a compromise is often more difficult, or even impossible, in situations when, for instance, the weld metal contains 0·6–1 % niobium, or more than 1 % silicon.

The foregoing explains the practical importance of the Schaeffler diagram, in allowing control of the composition of a filler metal so that its ferrite content is sufficient (generally above some %) to resist cracking at high temperature. This remedy is widely employed in 18 % Cr/10 % Ni and 24 % Cr/12 % Ni steels; it is not applicable to 25 % Cr/20 % Ni and 18 % Cr/38 % Ni steels, which do not contain ferrite by virtue of their composition, although very small amounts may be present in 25 % Cr/20 % Ni alloys. Nonetheless, it is possible to avoid cracking during the welding of these steels in several ways. One of them consists of using sufficiently pure filler metals which are free from known crack-promoting elements. Vacuum melting and the use of very pure raw materials for the metal preparation seem to improve the situation generally, probably by lowering the sulphur, phosphorus and oxygen contents.

Both these categories of refractory austenitic steels often contain more than 1 % silicon to improve their resistance to oxidation at elevated temperature. It is very difficult to avoid cracking when the weld metal itself contains more than 1 % silicon. As an example, the cracking intensity of a restrained 25 % Cr/20 % Ni/0·1 % C weld is twice as high when it

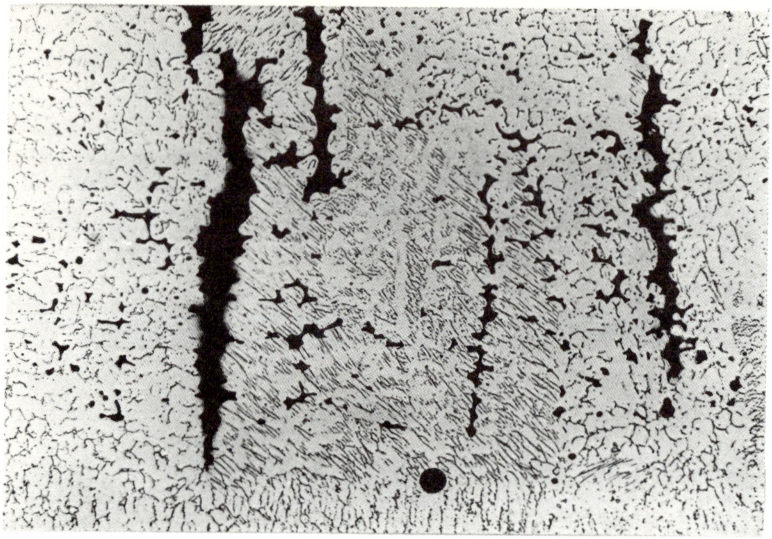

Fig. 6.11. Cracks in a weld containing about 6 % ferrite in a 19 % Cr/11 % Ni/0·8 % Nb steel (× 90).

contains 1·5 % silicon as when it contains only 0·2 % of this element. In this type of weld, the interdendritic regions are enriched in all alloying elements, as shown by the following results:

	Si	Mn	Ni	Cr
Dendrite axes	1·2	1·4	19	22·5
Interdendritic regions	2·8	2·7	22	27·5

The enrichment in nickel, manganese and, above all, in silicon undoubtedly leads to a reduction in the solidification temperature of the interdendritic liquid, thus allowing cracking to occur. Segregation can also lead to the formation of a distinct phase in the interdendritic regions as shown in fig. 6.12, which obviously plays a role in the mechanism of cracking, and may possibly be related to ferrite. This phase would have an unfavourable effect here, either because it is present in very small quantity and its solidification temperature is greatly reduced by its high silicon content, or because its mechanical strength is very low at elevated temperature.

In general, welds in 25 % Cr/20 % Ni and 18 % Cr/38 % Ni steels with a purity consistent with good industrial practice can be made without cracking, if the silicon content is below about 0·4 %.

It seems also that the ductility of these steels at elevated temperature and their resistance to cracking during welding are highest when the ratio

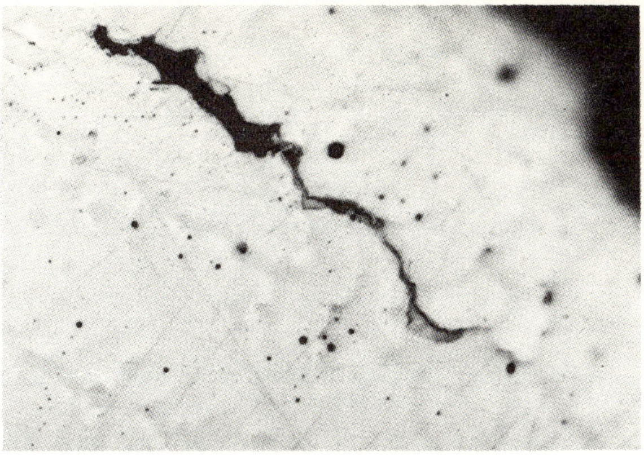

Fig. 6.12. Microcrack initiated at a discrete phase in the matrix in a 25 % Cr/20 % Ni/1·5 % Si steel weld metal (×900).

of their carbon and silicon contents is about 0·5. The cracking tendency appears to increase rapidly as this ratio decreases. It is possible that a rise in the carbon content compensates for the ferritising influence of silicon, and thus avoids the formation of interdendritic ferrite films which may liquate or have little resistance to cracking.

This proportion is easy to maintain in cast refractory steels, which contain a substantial quantity of carbon and silicon. In the case of wrought refractory steels, the carbon content is between 0·1 and 0·2 % and this necessitates control of the silicon content below approximately 0·4 %, as already pointed out above. However with 25 % Cr/20 % Ni steels intended to resist corrosion in aqueous media, the carbon content is often very low, and it is industrially impossible to lower the silicon sufficiently to maintain the ratio of 0·5.

Finally, there is a method based on the use of filler metals containing a high proportion of manganese. It has been shown practically that fully austenitic welds in all types of stainless steels are more resistant to cracking at elevated temperature when they contain 7–10 % manganese. The mechanism of this improvement is not well known, but the following hypotheses can be proposed:

– The presence of substantial amounts of manganese enables sulphur to be held as manganese sulphide, which melts at a higher temperature than does nickel sulphide. Moreover, manganese has a deoxidising effect.

– Manganese could increase the capacity for plastic deformation at grain boundaries at elevated temperature.

– At very high temperature, manganese could induce the formation of a second phase which disappears during cooling; as has already been pointed out, when the manganese content is above 4 %, it might act as a ferrite former at very high temperature, but without reducing the rate of transformation of ferrite into austenite.

4.5 Remarks

The presence of an easily melted phase is detrimental from the cracking point of view only when it is present in small amounts. This is the case with stainless steels stabilised by an addition of 0·6–1 % niobium, in which the formation of a small quantity of eutectic promotes cracking. If the amount of low melting point liquid is sufficient to fill cracks as they are formed until solidification is complete, cracking can be avoided. This is termed 'self-healing', and will be considered further in relation to nickel-base alloys containing 2 % niobium in chapter 7.

Procedures permitting the avoidance of cracking in the weld metal are also applicable to the HAZ, to the extent that the causes of cracking are similar. It is sufficient that the base metal has a ferritic tendency of a few %, or is sufficiently pure and free from crack-promoting additional or alloying elements, to avoid intergranular cracking in the zone affected during welding. When the ferritic tendency of the base metal is above 7–8 %, the welding thermal cycle causes the formation of small islands of delta ferrite in the grain boundaries of the affected zone adjacent to the fusion line; this can be seen in fig. 6.13, which shows the transition zone between the weld metal and base metal of an 18 % Cr/10 % Ni steel weld, with a ferritic tendency of approximately 15 %. The mechanism of this intergranular ferrite formation is similar to the mechanism of ferrite formation in the fusion zone, apart from the effect of segregation during solidification.

5 Corrosion resistance of welded assemblies

The HAZ and weld metal represent discontinuities in the structure and sometimes in the chemical composition of the weldment, and their presence can have repercussions upon the corrosion resistance of the assembly. The principal features of this important subject will be considered.

5.1 Galvanic corrosion

This type of corrosion is encountered in wet environmental conditions. It is well known that when two parts of a fabrication differ in composition, or at least in microstructure, and are immersed in an electrolyte, a galvanic cell is likely to be formed, with attack taking place on the anodic regions of the circuit. The attack is more severe as the galvanic current density is increased, i.e. as the surface area of the anodic regions is reduced in relation to the cathode area.

As the weld metal always has a much smaller surface than the rest of

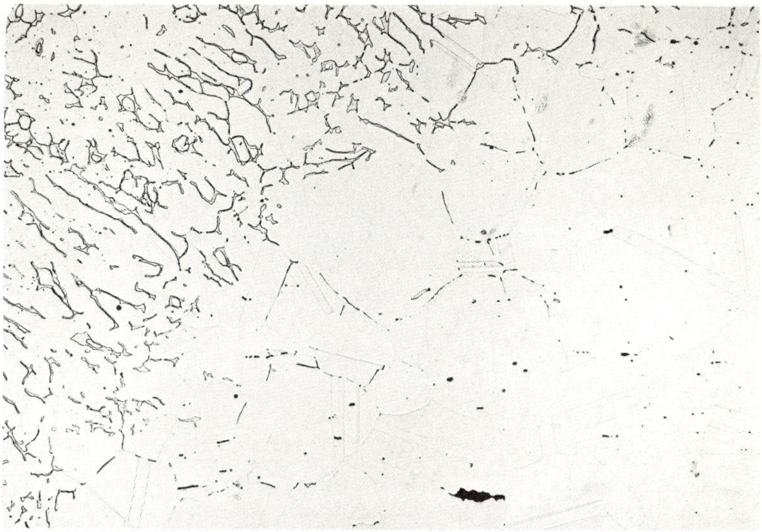

Fig. 6.13. Transition zone between weld metal and base metal in an 18 % Cr/10 % Ni steel weld with a ferritic tendency equal to 15 % (× 180).

the assembly, it is essential that it does not become anodic. To achieve this, the general principle is followed of employing a filler metal with a composition similar to that of the base metal, or with a slightly higher chromium content to compensate for any possible effect of the difference in microstructure. The electrochemical dissolution of the weld is thus suppressed, or transferred to the base metal, where its effect is reduced by the relatively large surface available.

In a weld having an austenite–ferrite structure, the differences in chemical composition between austenite and ferrite can promote the formation of local galvanic cells. This leads to a limitation on the ferrite content of the weld metal in some situations.

The presence of a weld can further cause the formation of galvanic cells indirectly. For example, the oxide layer formed on the weld and adjacent metal may render the weld anodic even though there is no variation in the chemical composition of the metal. This can result solely from the electrochemical properties of the oxide itself. Moreover, the presence of pores or cracks in the oxide can lead to the accumulation of corrosion products in the pores, or to fluctuations in the concentrations of aggressive agents and oxygen dissolved in the environment. Such chemical variations of the environment in contact with the metal can lead to local corrosion. Because of this, the oxide layer produced by a welding operation must be carefully removed by chemical or mechanical means before the weldment is put into service.

Geometrical irregularities of the weld such as undercut, or lack of

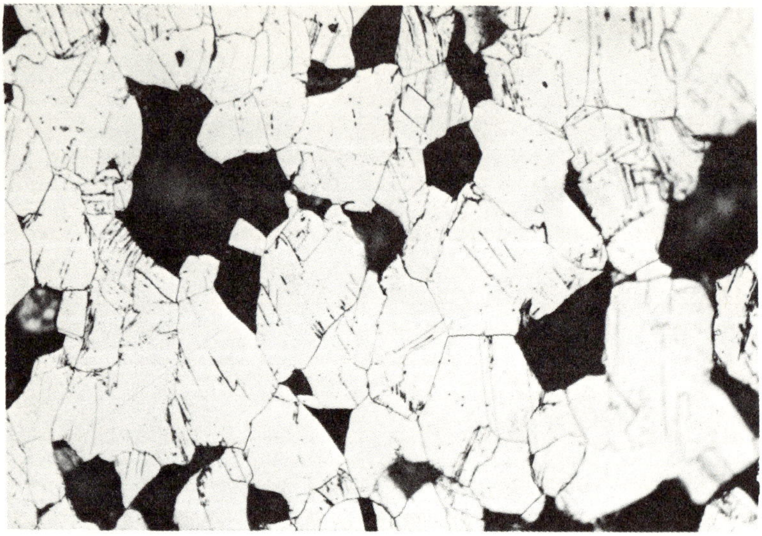

Fig. 6.14. Intercrystalline corrosion and dislodging of grains in an 18 % Cr/10 % Ni/0·06 % C steel (× 180).

penetration defects, can also produce crevices in which aggressive agents are not renewed, the dissolved oxygen diffuses slowly, and corrosion products can accumulate. These irregularities then become sites for preferential attack, and must be carefully avoided when making welds for corrosion-resistant applications.

5.2 Intercrystalline corrosion of welded assemblies

This phenomenon is well known, and only the generally accepted mechanism will be considered.

The carbon present in solution-treated austenitic stainless steels is retained in unstable solution at temperatures below approximately 900 °C. When these steels are subsequently heated in the temperature range between about 500 and 800 °C, such treatment being known as 'sensitisation', intergranular precipitation of chromium carbides can take place. The diffusion of chromium in austenitic steels is relatively slow at these temperatures and the precipitation leads to a depletion in chromium of the zones adjacent to the grain boundaries. This local impoverishment, if sufficiently pronounced, can cause a significant loss of corrosion resistance in numerous media, particularly acidic conditions such as sulphuric, sulphuric–nitric, hydrofluoric–nitric, hot nitric and hot organic acid environments. This phenomenon is called 'intercrystalline' or 'intergranular' corrosion, and is illustrated in fig. 6.14. The highly localised nature of the attack is evident, and this can dislodge the grains of the metal and lead to its rapid disintegration.

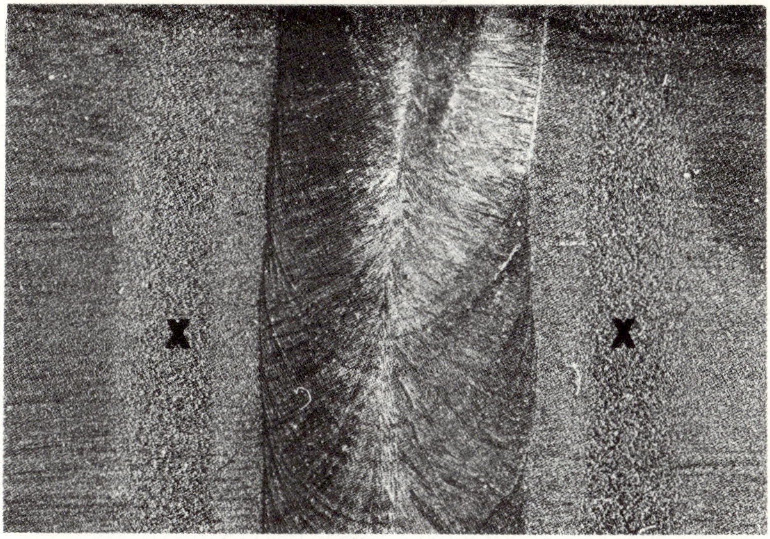

Fig. 6.15. Intercrystalline corrosion in the heat-affected zone of a weld in an 18 % Cr/10 % Ni/0·06 % C steel (× 2).

Susceptibility to intercrystalline corrosion depends on the composition of the material, and in particular on its carbon content. For a steel with a given composition, sensitisation can take place only under certain conditions of temperature and holding time. Sensitisation of 18 % Cr/10 % Ni steels in the range 550–800 °C requires longer times as the temperature is reduced, as would be expected from the laws of diffusion. At elevated temperature, sensitisation can be extremely rapid. For a specific temperature, it has been shown that the sensitisation resulting from heat treatment initially increases with the holding time, and then decreases. This can be explained on the basis that, above a certain amount of precipitation, an additional holding time causes the grain boundary carbides already formed to coalesce, and become discontinuous; further, and of greater importance, the chromium content will be equalised by diffusion. The recovery of corrosion resistance takes place more rapidly at higher temperatures within the sensitising range. The removal of sensitisation by prolonged holding at temperature has been sometimes called 'stabilisation'.

In a welded assembly, sensitisation can occur during welding in a zone situated at a certain distance from the weld bead, which is heated into the sensitising temperature range for a sufficient length of time. Between this zone and the weld metal, there is a region raised to higher temperatures, where carbides do not precipitate, or where the chromium diffusion is sufficiently fast for local depletion to be avoided. Fig. 6.15 shows a weld in an 18 % Cr/10 % Ni/0·06 % C steel which has been subjected to corrosion in acidic conditions, and has suffered intercrystalline attack in the sensitised regions (marked by a cross).

Although the concept that sensitisation is due to diffusion and precipitation of grain boundary carbides adequately accounts for the effects of relatively long holding periods, difficulty arises in applying this theory to welded joints. Experience has shown that sensitisation during welding can take place in extremely short times, not exceeding one or two seconds. Taking into account the composition of the chromium carbides, this rapid sensitisation would require an extremely high diffusion rate. In fact, in an 18 % Cr/10 % Ni/0·08 % C steel, diffusion would have to give local concentrations of carbon in one second, which are 10–100 times higher than that of the matrix. To remove this difficulty, it has been hypothesised and actually shown [19] that an enrichment of the grain boundaries in carbon takes place during the prior heating to very high temperature, for instance above 950 °C. As this enrichment occurs either during original solution treatment of the metal, or during the thermal cycle of the welding operation, it accounts for the fact that sensitisation can subsequently be produced by even a very short holding time between 550 and 800 °C.

Intercrystalline corrosion of welded assemblies can be avoided in many ways:

When the metal composition is such that sensitisation takes place during a welding operation, full corrosion resistance can be restored by treatment at a temperature between 900 and 1150 °C, followed by rapid cooling in air or water. This treatment removes the local chromium depletion by diffusion, when it is carried out at a relatively low temperature, and redissolves the precipitated carbides when it is carried out at higher temperatures. The rapid cooling avoids any further precipitation. It must be noted that, in contrast to the case of 17 % chromium steels (chapter 5, § 8.1), the temperature range in which immunity to attack is obtained in a reasonable time is situated above the sensitisation range. Consequently, it is impossible to restore the intercrystalline corrosion resistance of the assembly by carrying out a localised heat treatment of the weld, since this would have the effect only of displacing the sensitised zones a little further from the weld. It is therefore necessary to subject the whole assembly to heat treatment. This is often impossible and always inconvenient. Hence, physico-chemical means are generally employed, as they permit the risk of intergranular corrosion to be eliminated from the outset, without resorting to heat treatment after welding.

The most widely employed method is based on the use of austenitic stainless steels with a low enough carbon content for carbides not to precipitate in sufficient quantity to cause sensitisation during welding. For practical purposes, it is sufficient with 18 % Cr/10 % Ni steels that the carbon content is 0·04 % or below to avoid sensitisation when welding material of up to about 5 mm (0·2 in.) thickness. With thicker material, it is preferable to lower the carbon content to 0·03 % in order that the steel can be welded without risk of intercrystalline corrosion. These carbon contents can be slightly increased if the chromium content is 20 %, all other conditions remaining the same. An increase in the nickel content has an opposite effect, so that the limitations on carbon content are nearly the same for 25 % Cr/20 % Ni steels as for 18 % Cr/10 % Ni alloys. Steels of the 18 % Cr/10 % Ni type having a carbon content of 0·03 % max are in current use, and customers can now obtain material from steel-makers with a carbon content not exceeding 0·02 %, which remains completely immune to intercrystalline corrosion after the most severe welding operations on thick plates or after severe hot-working operations.

Another approach consists of using stainless steels with an average or low carbon content, which are 'stabilised' by an addition of titanium or niobium. Both these elements have the property of forming virtually insoluble carbides at elevated temperature, and if they are present in sufficient quantity in the metal, most of the carbon is tied up as titanium or niobium carbides or carbonitrides. Thus, carbon is no longer available to form chromium carbides in the temperature range over which sensitisation of unstabilised steels takes place.

The quantities of titanium or niobium to be used depend upon the difference between the carbon content of the steel and the carbon content remaining in solution at the sensitisation temperature, upon the composition of the titanium or niobium carbides which are formed, and upon the solubility of these two elements in austenite. In practice, titanium contents 4 or 5 times more than the carbon content, or niobium contents 8–10 times more, allow elimination of the risk of sensitisation by welding or hot-working.

It has been shown that cold-working improves resistance to intercrystalline corrosion after sensitisation by isothermal heat treatment at between 550 and 800 °C. This can be explained by the fact that cold-working multiplies the sites where carbide precipitation is possible, such as slip planes and twins. All other factors remaining constant, the localised chromium depletion is less marked, while stabilisation by chromium diffusion is easier and more rapid. Cold-working has the effect of reducing the duration of heat treatment at a given temperature, or the temperature range at a constant time, over which sensitisation can take place.

The situation regarding the HAZ of a weld in previously cold-worked material is more complicated, as it involves a very wide range of temperatures and heating times. In regions heated to a temperature below the recrystallisation temperature, the cold-working seems to reduce the width of the sensitised zone, by a mechanism similar to that causing the improvement found on isothermal sensitisation. In the regions heated above the recrystallisation temperature, some carbides precipitate at the grain boundaries, twins and slip planes during the initial heating period of the

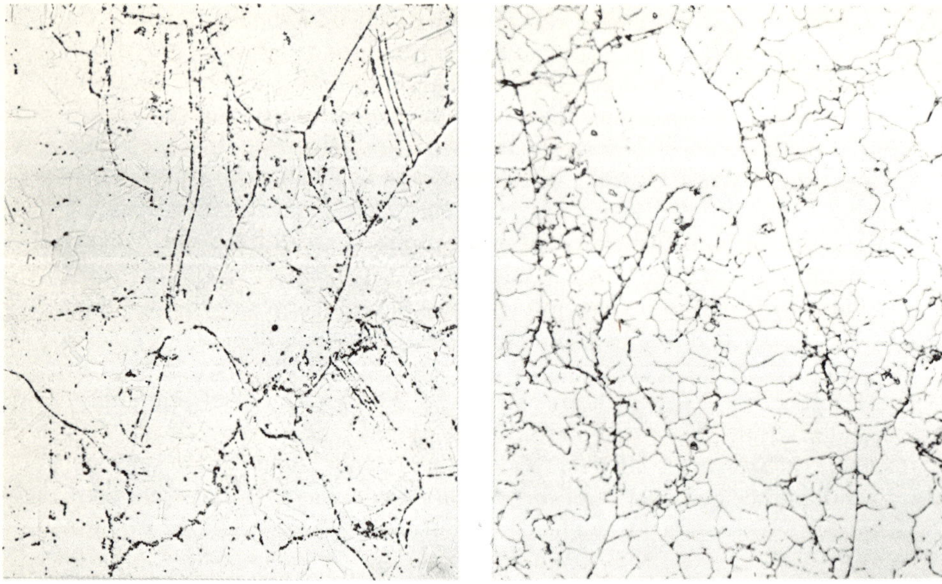

(*a*) (*b*)

Fig. 6.16. Heat-affected zone of a weld in a previously cold rolled 18 % Cr/10 % Ni/0·07 % C steel (× 400). (*a*) electrolytic chromic acid attack; (*b*) attack in a copper sulphate–sulphuric acid reagent after 24 h.

welding thermal cycle. The metal then recrystallises, and if the carbon content is sufficient, carbides can precipitate at the boundaries of the recrystallised structure during cooling or during deposition of any further weld runs. Fig. 6.16 gives an example of this behaviour in the HAZ of a two-pass weld in 3 mm (0·12 in.) thick cold rolled steel plate containing 18 % chromium, 10 % nickel and 0·07 % carbon. In fig. 6.16(*a*) the micrograph, obtained after an electrolytic chromic acid etch, shows the intense precipitation which has taken place prior to recrystallisation, even though practically no carbides are visible at the boundaries of the final recrystallised, fine grain microstructure. However, after immersion in a boiling acid copper sulphate reagent for 24 h (fig. 6.16(*b*)), the grain boundaries of the recrystallised structure are much more severely attacked than the slip planes or twins of the previous cold-worked structure. This example shows that the conclusions derived from studies on the influence of cold-working during isothermal sensitisation should not be extrapolated to the case of welded joints without considerable caution.

General Comments. A risk of intercrystalline corrosion may exist in the weld metal, either at the intersection of welds or when the weld is made using a number of runs. In the latter case, it is prudent to put only the last run deposited into contact with the corrosive environment, as this run has not been subjected to a sensitising treatment.

Nevertheless, it is normally found for a given composition that the austenite–ferrite structures obtained by welding are much less sensitive to intercrystalline corrosion, than solution-treated fully austenitic material. This can result from the increased number of precipitation sites associated with the presence of a second phase, from the relatively high chromium content of ferrite, or from the higher diffusion rate of chromium in this phase. The practical consequence is that restrictions on carbon content can be less severe for the weld metal than for the base material. The good behaviour of austenite–ferrite microstructures is further improved by the addition of 2–3 % molybdenum, which can render an austenite–ferrite weld containing 0·08 % carbon insensitive to intercrystalline corrosion. It should be noted that in molybdenum steels such as type 316 L, which are generally employed in aggressive acid media, partial transformation to sigma phase may take place if the ferrite is relatively slowly cooled as in submerged-arc welding. This transformation occurs without long range diffusion, and starts at the austenite–ferrite interfaces. The transformed regions are an aggregate of fine sigma and austenite particles, with the same total composition as that of the initial ferrite. In particular, there is no chromium-depleted zone bordering on these regions, and although, as a whole, they are more sensitive to corrosion than the adjacent austenite, especially in hot nitric acid, this corrosion has neither the same nature nor the same embrittling effects as the conventional intercrystalline corrosion [20].

5.3 'Knife-line' attack of welds in stabilised steels

This type of corrosion has the following distinctive features:

- It occurs only in steels of average or high carbon content which are stabilised with respect to intercrystalline corrosion by the addition of titanium or niobium.
- It arises only in hot concentrated nitric acid media.
- It affects only a narrow zone of the base metal immediately adjacent to the weld metal (fig. 6.17 shows an example of this type of corrosion).
- Carbon is detrimental in causing this type of attack as with intercrystalline corrosion but, in contrast to the situation with this latter problem, the corrosion resistance is not restored by heat treatment at elevated temperature.

The localised nature of this corrosion has led to its mechanism being sought in the structural phenomena which take place during heating to very high temperature approaching the liquidus–solidus interval, followed by rapid cooling. Recent investigations [21], [22] have enabled the following explanation to be proposed:

- During heating to above 1300 °C approximately, the grain boundaries of the base metal start to melt, causing the formation of a liquid phase

Fig. 6.17. Knife-line attack of a weld in a titanium-stabilised 18 % Cr/10 % Ni steel (× 3).

Fig. 6.18 Eutectiform carbide in a titanium-stabilised 18 % Cr/10 % Ni steel heated above 1300 °C (× 1600).

which is highly enriched in carbon and titanium (or niobium). On rapid cooling, this enrichment is retained at grain boundaries, and a eutectiform precipitate is formed, almost certainly at high temperature. An example of this is shown in fig. 6.18, obtained by electron microscopic examination of an extraction replica. Electron diffraction shows that this precipitate consists of titanium carbide, containing small quantities of chromium, iron, and perhaps nitrogen, in solution. Experience has shown that this carbide is attacked by nitric acid, the attack probably being facilitated by its branched morphology.

Thus, knife-line corrosion can be explained by dissolution of this carbide, the attack being subsequently continued in the cavities created as a result of the difficulties of renewal of the aggressive medium, and of its impoverishment in oxygen.

The mechanism of knife-line attack is therefore entirely different from that of intercrystalline corrosion, and accounts for the impossibility of restoring the corrosion resistance by heat treatment, as well as the specific nature of the environmental conditions causing corrosion. In practice, austenitic stainless steels with titanium or niobium must not be employed for welded fabrications in hot concentrated nitric acid media.

5.4 Stress corrosion

Stress corrosion is a type of attack affecting austenitic stainless steels of commercial purity which are subjected simultaneously to a tensile stress (and not a compressive stress) and to attack by chloride solutions. In these steels, attack is transgranular and can lead to the rapid destruction of the metal by cracking, with little weight loss of the material,

This type of corrosion is not specific to welded assemblies. However, welding can play an indirect role by creating stresses which increase the risk of attack, since they are additional to the externally applied stresses, or by producing a zone with different stress corrosion behaviour from that of the parent material.

The main characteristics of this type of corrosion are as follows:

- There is an incubation period during which no crack is visible; cracks then develop quite rapidly.
- Oxygen accelerates stress corrosion.
- The rate of attack is greatly reduced when the nickel content is above 40 % or is below 5 %.
- High purity metals are more resistant to stress corrosion.
- Special 18 % Cr/14 % Ni type compositions with more than 3·3 % silicon show a marked improvement in the conventional magnesium chloride laboratory test [23], although the use of this medium to reproduce behaviour in other chlorides has been questioned.
- The surface condition plays an important part. Sand or shot blasting improves the resistance, possibly by creating compressive stresses at the surface. On the other hand, finishing procedures which are liable to cause tensile surface stresses must be prohibited. The bright annealing of plates has a favourable effect upon their stress corrosion resistance.
- The metal structure plays an important role. Austenite–ferrite alloys have a better resistance than fully austenitic materials. An improvement is

also noted when a high proportion of austenite has been transformed into martensite by cold-working of the entire fabrication.

Provided that they contain a substantial quantity of ferrite, 18 % Cr/10 % Ni steel weld metals are more resistant to stress corrosion than the base metal.

6 Mechanical properties of austenitic stainless steel welded assemblies at room temperature

6.1 Mechanical characteristics of weld metals

Generally, austenitic stainless steel weld metals in the as-welded condition have a tensile strength comparable with that of the solution-treated and quenched parent material, but a significantly higher strength at low deformations. This superiority at low applied strain has two main causes:

– The weld metal is the site of significant residual stresses. This pre-existing stress can act either directly, or indirectly by causing plastic deformation of the metal. It is well known that very light prior plastic deformation considerably increases the strength of austenitic stainless steels at small applied strains.

– The ferrite which appears to some extent in 18 % Cr/10 % Ni and 24 % Cr/12 % Ni type weld metals has a hardening effect. This effect is particularly marked at low strains, but it also raises the tensile strength.

This good mechanical strength of welds is accompanied by a reduction in toughness at room temperature. For example, in 18 % Cr/10 % Ni and 18 % Cr/10 % Ni/Mo steels, the UF impact value of solution-treated wrought material exceeds 150 J/100 mm^2, whereas that of the weld deposits is usually between 60 and 120 J/100 mm^2. If the weld deposit is solution treated and rapidly cooled, the ferrite largely disappears, while most of the internal stresses are relieved. This treatment results in an increase in the impact strength of the deposit, and renders its tensile properties more comparable with those of the parent material.

To illustrate the foregoing, table 6.1 gives some average mechanical properties obtained for various types of filler metal using all-weld metal testpieces. The strength at low strain is represented by the conventional 0·2 % proof stress $E_{0 \cdot 2}$. For each filler metal, the average ferrite contents of the deposits on which the tests have been carried out are indicated. It is apparent that for 18 % Cr/10 % Ni and 18 % Cr/10 % Ni/Mo materials, the tensile strength and 0·2 % proof stress can vary by approximately ± 50 N/mm^2 (3 tonf/$in.^2$) when the ferrite content is below 1 % or above 20 %.

It should be noted that a high ferrite content (29 % Cr/9 % Ni type steel) enables a high strength to be obtained, but a low toughness as a result of the notch sensitivity of chromium ferrites discussed in chapter 5. In this

Table 6.1. *Mechanical characteristics of weld deposits*

Weld metal	Treatment	Tensile strength, R [N/mm²(tonf/in.²)]	$E_{0\cdot2}$ [N/mm² (tonf/in.²)]	Elongation (%)	Toughness UF, [J/100 mm²]
18 % Cr/10 % Ni C = 0·04 % 8 % ferrite	As-deposited	580 (38)	420 (27)	35	⩾ 60
	1050 °C	550 (36)	210 (14)	45	⩾ 90
18 % Cr/10 % Ni/ 2 % Mo C = 0·05 % 8 % ferrite	As-deposited	660 (43)	490 (32)	30	⩾ 60
	1050 °C	580 (38)	260 (17)	40	⩾ 90
24 % Cr/12 % Ni C = 0·12 % 2 % ferrite	As-deposited	640 (41)	480 (31)	30	⩾ 45
	1050 °C	580 (38)	300 (19)	40	⩾ 70
25 % Cr/20 % Ni C = 0·08 % 0 % ferrite	As-deposited	600 (39)	400 (26)	25	⩾ 50
	1050 °C	570 (37)	270 (18)	35	⩾ 65
29 % Cr/9 % Ni C = 0·12 % 50 % ferrite	As-deposited	860 (56)	670 (43)	20	⩾ 30
	1100 °C†	780 (50)	530 (34)	25	⩾ 35

† Water-quenched. Others air-cooled.

case, rapid cooling from 1100 °C does not reduce the ferrite content sufficiently to give much improvement in the impact strength.

6.2 Mechanical properties of welded assemblies

From the mechanical point of view, a welded joint is composed of three different regions:

- The metal unaffected by heat.
- The weld metal, which has a strength superior or equal to that of the unaffected metal.
- The HAZ; in this zone, the metal adjacent to the fusion line has suffered some grain growth, which slightly reduces its mechanical strength, particularly at low strains.

It could therefore be expected that tensile testpieces taken transverse to a welded joint would consistently break in the HAZ, and that the strength would be lower than that of a testpiece taken from the parent material alone. This is not usually the case, for the following reasons:

- The softer region of the HAZ is generally narrow.
- This softer region is restrained on both sides by harder material which opposes its deformation.

Table 6.2. *Transverse mechanical properties of welded joints in* 3 *mm* (0·12 *in.*) *thick plates*

Welding with covered electrode having a composition similar to that of the base metal. Fractures in the base metal or in the HAZ.

	Parent material			Welded joint		
	Tensile strength [N/mm² (tonf/in.²)]	$E_{0\cdot2}$ [N/mm² (tonf/in.²)]	Elongation (%)	Tensile strength [N/mm² (tonf/in.²)]	$E_{0\cdot2}$ [N/mm² (tonf/in.²)]	Elongation (%)
18 % Cr/10 % Ni (C = 0·025 %)	580 (38)	250 (16)	57	585 (38)	260 (17)	44
18 % Cr/12 % Ni/2 % Mo (C = 0·025 %)	605 (39)	320 (21)	48	610 (40)	330 (21)	39
25 % Cr/20 % Ni (C = 0·10 %)	710 (46)	380 (25)	39	725 (47)	400 (25)	33

Table 6.3. *Transverse mechanical properties of welded joints in* 2 mm (0·08 in.) *thick plates*

Automatic TIG welding without filler metal. Fracture along the weld axis.

	Parent material			Welded joint		
	Tensile strength [N/mm² (tonf/in.²)]	$E_{0\cdot2}$ [N/mm² (tonf/in.²)]	Elongation (%)	Tensile strength [N/mm² (tonf/in.²)]	$E_{0\cdot2}$ [N/mm² (tonf/in.²)]	Elongation (%)
18 % Cr/10 % Ni (C = 0·025 %)	570 (37)	280 (18)	52	580 (38)	275 (18)	44
18 % Cr/12 % Ni/2 % Mo (C = 0·025 %)	600 (39)	300 (19)	47	595 (39)	320 (21)	33
25 % Cr/20 % Ni (C = 0·10 %)	700 (45)	430 (27)	38	700 (45)	410 (27)	28

– Preferential deformation in this zone at the start of the tensile tests has a work-hardening effect, so that the strength of the remainder of the testpiece is quickly attained.

In practice, it is found that fracture of such a transverse testpiece takes place in the HAZ, or in the unaffected base metal when the cross-sectional area of the weld bead is greater than that of the base metal. The mechanical strength of the joint is then equal or superior to that of the parent material, as shown by the results in table 6.2.

In automatic TIG welds in thin plates made autogenously, a groove is frequently formed in the centre of the weld metal at the junction of solidification fronts. Fracture then occurs along the weld axis. Even in this case, provided that this depression is not too deep, the mechanical characteristics of the assembly remain similar to those of the base metal, as illustrated in table 6.3.

Austenitic stainless steel plates with high mechanical characteristics can readily be obtained by cold-working. A welding operation performed on these plates leads to the recrystallisation of the metal in part of the HAZ. These recrystallised regions have a hardness comparable to that of the solution-treated product, and thus below the hardness of the metal unaffected by the welding heat. Despite the presence of these soft regions, the mechanical characteristics of the joint may be considerably higher than those of the solution-treated base metal, for reasons considered above. Fig. 6.19 and 6.20 give examples of mechanical characteristics of testpieces taken transversally to cold rolled 18 % Cr/10 % Ni and 18 % Cr/10 % Ni/Mo plates welded by the TIG process without filler metal, and by the manual metal-arc process [24]. In these figures, the ratio of the tensile property (R or $E_{0 \cdot 2}$) of the joint to that of the plate in the solution-treated condition has been plotted as a function of the cold-working of the plate, as defined by the ratio between the cold-worked plate property and that of the same plate in the solution-treated condition. Each point is the average of six tests, and envelopes are drawn to show the range of material behaviour. These graphs clearly show that it would not be logical to consider the mechanical strength of the welded joint as equal only to that of the solution-treated base metal. They also show that, with light deformation of the base plate, as normally results from planishing or polishing of the solution-treated plates (the groups of points situated close to the origin in figs. 6.19 and 6.20), the mechanical characteristics of the joints are equal or superior to those of the base metal. In consequence, provided that the welds do not contain defects liable to impair their mechanical strength, the adoption of a joint efficiency of 100 % is justified for this type of material.

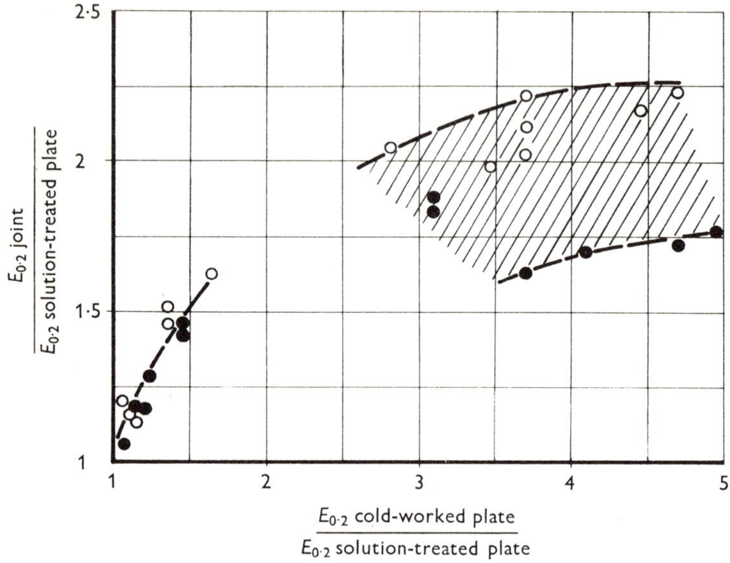

Fig. 6.19. 0·2 % proof stress $E_{0·2}$ of welded joints in cold worked 18 % Cr/10 % Ni steel plates: ● TIG welding without filler metal; ○ welding with covered electrode.

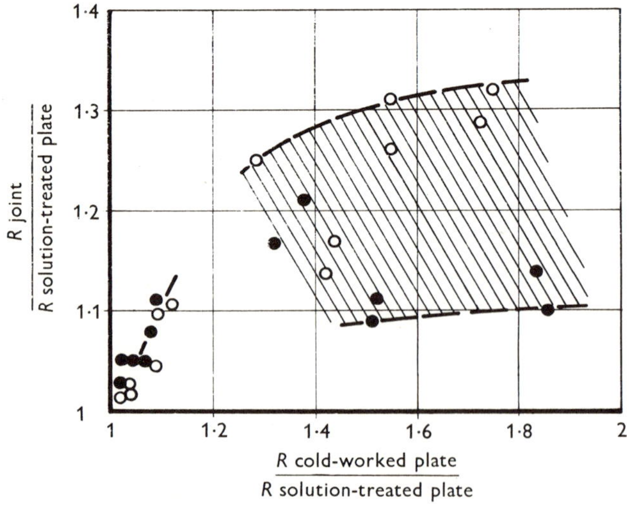

Fig. 6.20. Tensile strength R of welded joints in cold worked 18 % Cr/10 % Ni steel plates: ● TIG welding without filler metal; ○ welding with covered electrode.

7 Low temperature mechanical properties of joints in austenitic stainless steels

In contrast to martensitic and ferritic steels, 18 % Cr/10 % Ni and more highly alloyed types of austenitic stainless steels are not brittle at very low temperatures, as long as the austenite grain boundaries do not contain precipitated carbides. The toughness of these steels diminishes only very slowly as the temperature decreases, and their UF impact value remains above 100 J/100 mm^2 at the temperature of liquid nitrogen. Even the less alloyed types, which are partially transformed into martensite by cooling below room temperature, retain an acceptable toughness at very low temperature.

A decrease in temperature causes a slight increase in the proof stress of 18 % Cr/10 % Ni type steels, and a considerable increase in their tensile strength, probably because of the formation of martensite during a tensile test at very low temperature.

Austenitic stainless steel welds have the same properties as the base metal from this point of view, provided that the weld metal does not contain too much ferrite. Indeed, the impact transition temperature of the delta ferrite present in the weld beads is not much below room temperature. As long as the amount and distribution of ferrite in the structure are such that it is found in the form of individual islands, the presence of ferrite is not detrimental to the toughness of the weld metal, and the impact strength at −196 °C will be only slightly lower than at room temperature. However, as soon as ferrite is present in sufficient quantity to form a continuous network, it creates a path for brittle fracture at low temperature. This is illustrated by fig. 6.21, in which the Charpy V impact energy at −196 °C of 20 % Cr/14 % Ni/3 % Mo type weld metals containing 0·05 % carbon has been plotted against the ferrite content. The welds were made by the TIG process in 15 mm (0·6 in.) thick plates. The negative values of the ferrite content were obtained by application of the Séférian formula (§2.2), and simply show that the weld metal composition is some distance from the critical composition at which ferrite is likely to be formed.

In this example, a rapid drop in the impact value can be observed above a ferrite content of 12 %. The thermal conditions of welding influence the quantity and distribution of ferrite, and it is possible for a continuous network to be formed at a lower ferrite content. Thus in practice, it is advisable to limit the ferrite content of filler metals to 8 % for weldments intended to operate at very low temperature.

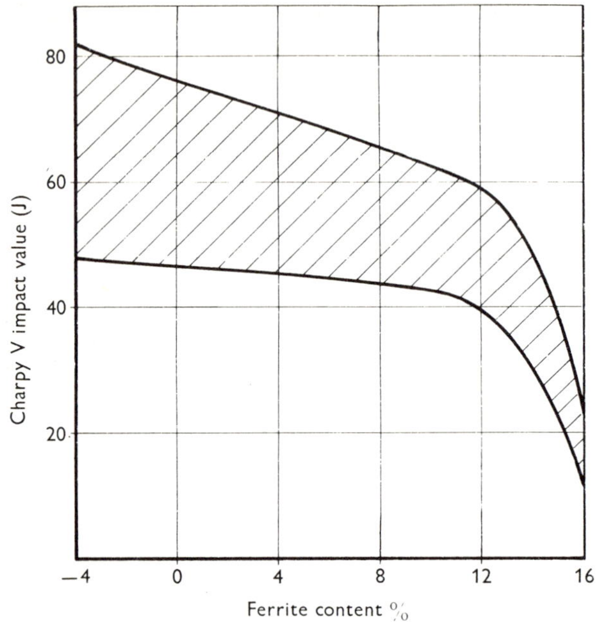

Fig. 6.21. Impact value at −196 °C of 20 % Cr/14 % Ni type weld deposits with 3 % molybdenum, as a function of their ferrite content.

8 Elevated temperature mechanical properties of austenitic stainless steel weldments

Apart from their resistance to corrosion at elevated temperature, austenitic stainless steels usually have good high temperature mechanical strength. Because of this property, they are extensively employed for thermal applications in which the working temperatures are too high for martensitic steels to be satisfactory. In general, the steels used for their high temperature mechanical properties are derived from the 18 % Cr/10 % Ni type, with moderate carbon content and some additional elements intended to improve their creep resistance (niobium, titanium, molybdenum, tungsten, copper etc.). More highly alloyed grades are also available.

In addition to the general necessity for good service behaviour, notably from the point of view of cracking during welding, welds meant for elevated temperature service must have a low susceptibility to embrittlement on prolonged holding at high temperature, and must possess a creep resistance similar to that of the base metal. These two problems will be considered successively.

8.1 Embrittlement by formation of sigma type intermetallic compounds

It was pointed out in chapter 5, §9, that, in iron-chromium alloys, there exists a range of temperature and composition centred on 45 % chromium, in which an intermetallic compound known as 'sigma phase' is likely to be formed. Similar intermetallic compounds can be formed in austenitic stainless and heat-resisting steels and in their welds and, although they may contain high proportions of elements other than iron and chromium, they are designated by the general term sigma phase.

Sigma phases are hard and brittle, and if present in significant amounts, they have the effect of reducing the ductility of the metal (i.e. the elongation at fracture) and the toughness, and of increasing its hardness. Although the importance of toughness in the context of high temperature service must not be exaggerated, these consequences are considered deleterious, and it is desirable to limit the formation of sigma phase in austenitic steels and in welded joints in these materials.

For an intermetallic compound of the sigma type to be formed in an austenitic steel weld, certain thermal, compositional and microstructural conditions must be combined.

Thermal conditions. In the as-welded condition, 18 % Cr/10 % Ni, 24 % Cr/12 % Ni or 25 % Cr/20 % Ni type stainless steel welds do not normally contain any sigma phase, because the rapid cooling retains the structure existing at very high temperature where sigma phase does not form. To cause sigma formation, the steel must be maintained at a temperature below the upper limit of sigma stability, which is generally around 900–950 °C.

Further, the formation of sigma type phases requires the diffusion of a number of elements, and in practice it can take place only at temperatures where compositional and structural conditions are such that these elements diffuse at an appreciable rate, i.e. above about 550 °C. It should be recognised that transformation is extremely slow between 550 and 600 °C, with the result that exposure periods of thousands of hours, or even of several ten thousands of hours in this temperature range are necessary for the mechanical consequences of sigma formation to become apparent. In general, transformation to sigma type phases is most rapid towards 800–850 °C, temperatures at which the diffusion rates are high, and where the compositional range over which these phases exist is still quite wide.

Compositional conditions. It is known that sigma phase formation is favoured by the presence of a high proportion of ferrite formers in the steel. For instance, the local chromium content must exceed 20 % for a sigma type phase to be formed in an austenitic stainless steel weld containing no other ferrite formers. An increase in the proportion of ferrite formers tends to increase the quantity of sigma phase likely to be formed, and to accelerate its formation. High silicon and molybdenum contents have the additional effect of increasing the temperature range over which sigma phase exists; as an example, a steel containing 20 % chromium, 10 % nickel and 3 % molybdenum still forms a sigma phase at 1000 °C, whereas in molybdenum-free compositions, this phase ceases to exist above about 900 °C. The effect of silicon has less practical importance, since the presence of significant amounts of silicon in welded joints is undesirable because of its adverse influence upon susceptibility to cracking during welding. However, welds in 24 % Cr/12 % Ni and 25 % Cr/20 % Ni steels made without filler metal, or with a filler metal having the same composition as that of the base metal, may contain more than 1 % of this element. All ferrite formers are present in high amounts in sigma type intermetallic compounds, so that diffusion rates of individual elements play an important role in their formation.

On the other hand, high carbon and nickel contents diminish the tendency for standard stainless steels to form sigma type phases. Carbon acts mainly by forming complex carbides of chromium, molybdenum and tungsten, which renders these elements unavailable to form other phases, while an increase in the nickel content has the effect of retarding the diffusion of ferrite formers in the austenite, and consequently slows down the rate of precipitation of intermetallic compounds.

It should be noted that manganese has the same influence on sigma formation as do ferrite formers.

Tables 6.4 (forged steels) and 6.5 (weld metal deposited by the TIG process) illustrate the foregoing, by some examples related to 25 % Cr/ 20 % Ni steels; the UF impact value after 1000 h of heat treatment at 800 °C has been chosen as a criterion of embrittlement by precipitation of sigma type phase. The UF impact value of the forged steels in the solution-treated condition is between 150 and 250 J/100 mm^2, while that of the as-deposited weld metal is between 90 and 120 J/100 mm^2 with the exception of deposits containing niobium, which have an impact value of 60 J/ 100 mm^2 in the as-welded condition. The results of table 6.4 show:

- The effect of chromium, by comparison of alloys 1 and 2.
- The influence of silicon, in spite of the difference in carbon contents, by comparison of alloys 1 and 3.

The results of table 6.5 illustrate the effects of manganese (cf. alloys 1–4) niobium (cf. alloys 3 and 7) and carbon (cf. alloys 2, 3, 5 and 6).

Figure 6.22 shows the metallographic appearance of sigma phase precipitation in alloys 1 and 4 of table 6.5, which are fully austenitic in the as-welded condition. The precipitated particles are stained by etching, and appear as irregularly shaped islands at the grain boundaries, and in acicular form in the interior of the grains.

Sigma type phases are not formed in the refractory alloys containing 18 % chromium, 38 % nickel and 0·10 % carbon; two forged alloys of this type, with 0·45 % and 2·1 % silicon, have UF impact values of 160 and

Table 6.4. *Embrittlement by sigma phase formation in forged and solution treated* 25 % Cr/20 % Ni *alloys*

Alloy	C	Si	Mn	Ni	Cr	UF impact energy after 1000 h at 800 °C (J/100 mm²)
1	0·03	0·36	1·8	19·5	24·5	58
2	0·07	0·40	1·8	20·0	26·3	18
3	0·16	1·7	1·4	19·3	25·0	10

Table 6.5. *Embrittlement by sigma phase formation in* 25 % Cr/20 % Ni *weld deposits*

Alloy	C	Si	Mn	Ni	Cr		UF impact energy after 1000 h at 800 °C (J/100 mm²)
1	0·06	0·36	1·9	19·5	24·9		28
2	0·05	0·42	6·2	19·0	24·0		22
3	0·05	0·43	9·4	19·5	23·9		13
4	0·05	0·50	11·5	19·0	24·0		4
5	0·025	0·52	6·2	19·5	24·0		18
6	0·025	0·53	9·2	19·5	23·5		11
7	0·06	0·56	9·0	19·5	23·5	Nb: 1·5	2

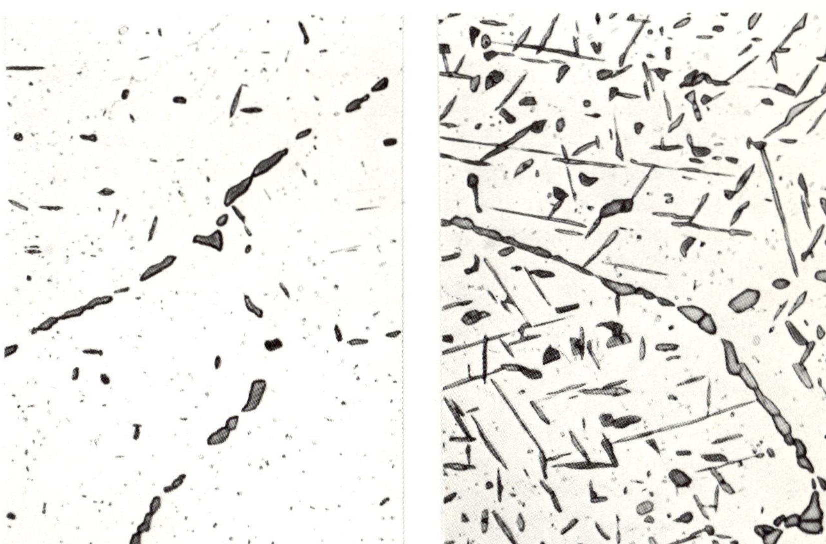

Welding 1 Welding 4

Fig. 6.22. Micrographs of welds 1 and 4 of table 6.5 after a heat treatment of 1000 h at 800 °C (× 350).

145 J/100 mm^2 respectively, after a heat treatment of 1000 h at 850 °C. However, silicon is not advisable in welds in this type of stainless steel, because of the associated risk of hot cracking.

Finally, it must be noted that the brittleness produced by sigma diminishes slightly when the temperature rises, but there is no transition temperature for this brittleness in the range over which sigma phase exists. For instance, the impact values at 20, 400 and 800 °C of a 25 % Cr/20 % Ni steel in which 15 % sigma phase has been precipitated by heat treatment of 300 h at 800 °C, are 14, 23 and 38 J/100 mm^2 respectively.

Structural conditions. Sigma phase can be formed in fully austenitic weldments when the thermal and compositional conditions are fulfilled, as shown in the examples of the preceding paragraphs.

However, the particular structure of welds can facilitate the formation of intermetallic compounds in steels having such a composition that these compounds would not appear in the same alloy when forged and quenched from the annealing temperature. Indeed, segregation during solidification can locally raise the concentration of ferrite formers, and make possible the formation of sigma type intermetallic compounds without the necessity for extensive diffusion. This is especially true in the case of austenitic–ferritic welds, because the delta ferrite is richer in ferrite formers than austenite; moreover, the diffusion rates are higher in ferrite than in austenite, so that transformation to sigma phases is more rapid when these phases are formed from ferrite than when they are formed from austenite. It is even possible that, when the weld deposit contains a high proportion of elements accelerating the formation of intermetallic compounds, ferrite present in the initial weld runs is transformed into sigma phase by reheating during deposition of subsequent passes. This is illustrated by the micrograph in fig. 6.23, which shows an 18 % Cr/13 % Ni steel weld with high silicon and manganese contents. The left-hand micrograph (Fig. 6.23(*a*)) refers to the last run performed, in which ferrite has not been transformed; fig. 6.23(*b*) shows one of the first runs, in which ferrite has been transformed into sigma phase and austenite by a eutectoid type decomposition; the corresponding aggregate is much more readily stained by etching than the original ferrite.

8.2 Practical consequences – choice of a filler metal for high temperature service

For practical purposes, the foregoing can be summarised as follows:

- 18 % Cr/38 % Ni steel welds do not form sigma phase.
- On prolonged heat treatment at between 600 and 900 °C, 25 % Cr/20 % Ni fully austenitic welds form some sigma phase from austenite. This is also the case with 24 % Cr/12 % Ni welds, although these often contain a little ferrite.
- On prolonged heat treatment, at between 600 and 900 °C, and above,

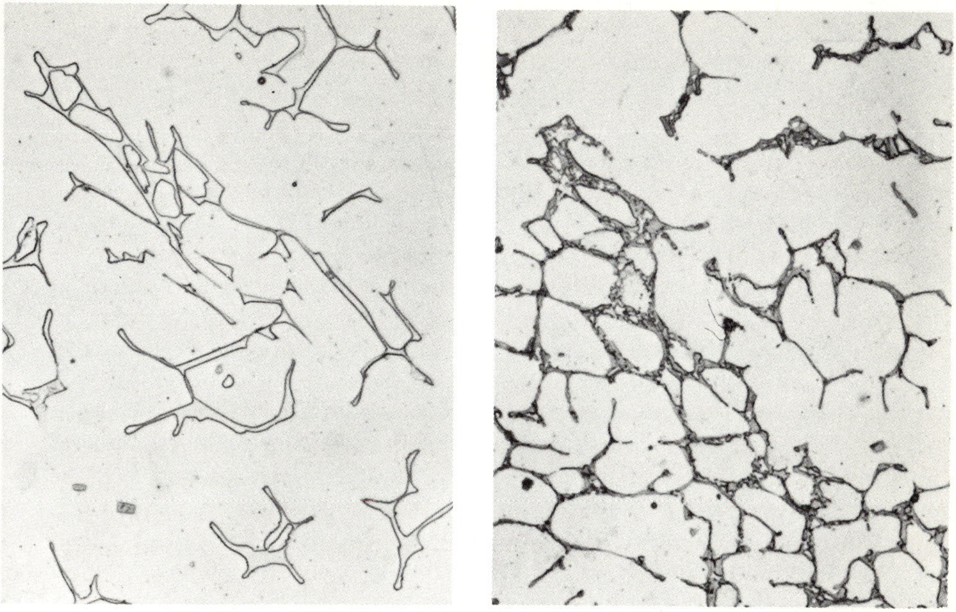

(*a*) (*b*)

Fig. 6.23. Structures of (*a*) final and (*b*) initial runs of a weld in an 18 % Cr/13 % Ni steel having high silicon and manganese contents (×400).

when certain additional elements are present, 18 % Cr/10 % Ni welds with a relatively high chromium content, or containing ferrite formers to improve their high temperature properties (molybdenum, tungsten, titanium, niobium etc.), form sigma phase by decomposition of delta ferrite.

The formation of these intermetallic compounds causes a loss of the notch toughness at all temperatures. A discussion of the real risk resulting from this falls beyond the scope of this book, but this risk must not be exaggerated as the untransformed austenite retains an excellent resistance to notch effects.

It is impossible to avoid sigma phase formation in 24 % Cr/12 % Ni and 25 % Cr/20 % Ni welds because of their high chromium content. Nevertheless, the quantity of this phase can be limited by avoiding filler metals which are rich in chromium, or contain excessive amounts of silicon, niobium or manganese.

In the case of 18 % Cr/10 % Ni welds, the embrittlement resulting from the decomposition of ferrite is important only when this phase is present initially in sufficient quantity to form a continuous network, as the sigma phase then creates a path for brittle fracture through this material. It is therefore recommended that only filler metals having a relatively low ferrite content be used for welds in heat-resisting steels. However, the presence of some ferrite is desirable to avoid cracking during welding, and a compromise between these requirements is necessary. This compromise generally involves a limitation on ferrite content, to within about 2–8 %.

As has already been pointed out in § 4.4, it can be difficult, or even impossible, to achieve such a compromise in the presence of elements which promote cracking during welding. This is the case with niobium-stabilised 18 % Cr/10 % Ni steels, which have a good creep resistance, but are critical when welded with a filler metal of matching composition to that of the parent material.

The risk of sigma embrittlement is often further reduced by keeping the chromium content of the deposited metal at a relatively low level, so that, allowing for elements intended to improve the creep resistance, only partial transformation of the ferrite occurs. As an example, this is achieved with the use of filler metals containing 15·5–16·5 % chromium, 7·5–8·5 % nickel, 0·05–0·10 % carbon and nearly 2 % molybdenum or about 4 % tungsten, the composition of the deposited metal being controlled so that it contains approximately 5 % delta ferrite. After heat treatment of 1000 h at 800 °C, this type of weld has an impact value of 70 J/100 mm^2, whereas the impact value of a weld with the same ferrite content but with 17·5 % chromium, 10·5 % nickel and 2·6 % molybdenum is no more than 30 J/100 mm^2.

These filler metals can be employed to weld the standard creep-resistant austenitic steels: 18 % Cr/10 % Ni/Ti, 18 % Cr/10 % Ni/Nb, 18 % Cr/10 % Ni/Mo, 18 % Cr/10 % Ni/Mo/Ti, 18 % Cr/10 % Ni/W/Ti. Owing to their relatively high carbon content, and to the presence of molybdenum or tungsten, the corresponding welds have high temperature mechanical properties similar to those of the base metal. Fig. 6.24 gives the times to fracture as a function of the initial stress at 650 and 700 °C for a steel containing 16·5 % chromium, 13·5 % nickel, 4 % tungsten, 0·08 % carbon and 0·5 % titanium (broken lines), and for testpieces taken transverse to welds made in this steel with a filler metal containing 15·6 % chromium, 8 % nickel, 3·5 % tungsten and 0·08 % carbon.

At high stresses, the times to rupture for welded joints are below those of the base metal, but these times become equal with durations of about 5000 h.

Consequently, it can be appreciated that the welding of creep-resistant 18 % Cr/10 % Ni steels is often carried out with a filler metal having a composition quite different from that of the parent material. This may be for reasons related either to weldability (as with 18 % Cr/10 % Ni/Nb steels), or to embrittlement by sigma phase. Alternatively, in certain welding situations it is difficult to ensure that elements giving good high temperature properties to the parent material are present in the deposited

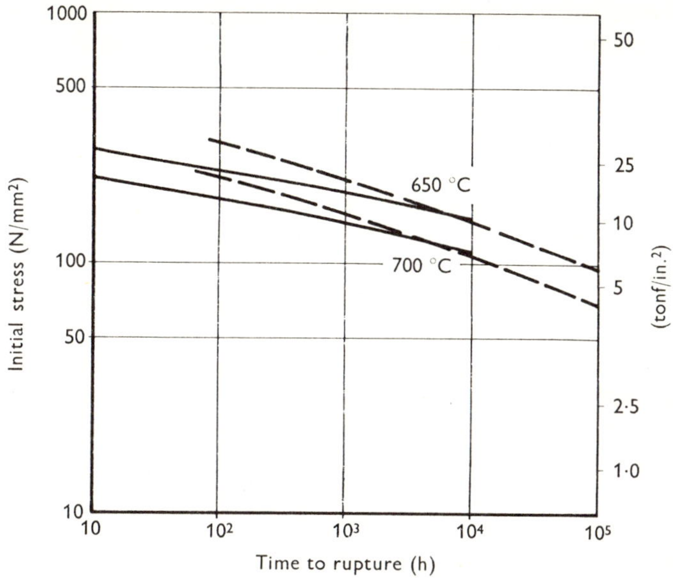

Fig. 6.24. Creep-rupture times as a function of the initial stress: ——— for a steel with 16·5 % Cr/13·5 % Ni/4 % W/0·5 % Ti; —— for welded joints in this steel made using a filler metal with 15·6 % Cr/8 % Ni/3·5 % W/0·08 % C.

weld metal. This is the case with 18 % Cr/10 % Ni/Ti steels for example, where titanium may readily be lost during transfer across the arc.

The situation is different for certain highly alloyed compositions with high mechanical characteristics at elevated temperature, since the use of a less alloyed filler metal for welding would not enable sufficiently strong joints to be obtained. This is, for instance, the case with alloys containing 20 % chromium, 20 % nickel, 20 % cobalt, 2·5 % molybdenum, 2·5 % tungsten, 1 % niobium and 0·15 % carbon; these materials must be welded with a filler metal of matching or more highly alloyed composition. The mechanical properties at high temperature of joints welded in this way are equal or superior to those of the base metal, but they are obtained at the cost of often considerable practical difficulties.

8.3 Stress-relieving of welds

Because of the low thermal conductivity, high coefficient of expansion and good elevated temperature mechanical strength of austenitic stainless steels, welds in thick material can be subject to high internal stresses in addition to the normal external stresses anticipated in service. However, owing to their excellent ductility, and their insensitivity to notches and stress concentrations, these steels have a remarkable property of adaptation to applied stresses, and do not generally require any stress-relieving treatment after welding.

It should also be noted that the high temperature mechanical properties of these steels render the low temperature treatments which are applied to ordinary or low alloy steels ineffective. For stress-relieving to be carried out efficiently in a reasonable time, it must be carried out at a sufficiently high temperature, such as above 800 °C, and this is often impossible in practice.

Nevertheless, heat treatment after welding can be useful with thick material, when the metallurgical phenomena caused by the rise in temperature and by the joint restraint reduce the capacity for plastic deformation of the HAZ. This is principally encountered in niobium-stabilised 18 % Cr/10 % Ni steels with a moderate carbon content, and the phenomena can be compared to those described by Hodierne for the heat treatment of drawn tubes in 18 % Cr/10 % Ni/Nb steels[25].

In the regions of metal heated to 1250 °C and above during welding, a general dissolution of all carbides takes place. Carbon can then reprecipitate largely in the form of intergranular chromium carbides on subsequent heating to temperatures above 550 °C. When this precipitation occurs under conditions of high stress (residual stresses and external stresses) and at a relatively low temperature, it is fine and continuous, and can lead to intergranular separation.

Other authors[26] have proposed a slightly different mechanism, in which the precipitation under stress after high temperature heat treatment takes place in a very fine intragranular form. Intergranular cracking then results from a general decrease in the ability of the metal to accomodate plastic deformation. It is possible that these two explanations are complementary.

In both cases, the preventative treatment consists of annealing the weld at 850 °C, before it is put into service and external stresses are applied. This treatment has the effect of reducing the magnitude of the internal stresses, and of precipitating the carbides in a coarse, non-dangerous form.

This type of failure due to carbide precipitation can also occur, though it seems much rarer, in other types of stabilised or unstabilised steels with a medium carbon content. As an example, fig. 6.25 shows an intergranular fracture in the HAZ of a weld in an 18 % Cr/10 % Ni/Mo/Ti/0·10 % C steel. Failure occurred after 20000 h service at 550 °C under an external stress, and was due to a very slow precipitation of chromium carbides at the grain boundaries.

Heat treatment at 850 °C prior to service could permit this type of fracture to be eliminated systematically, but the best solution seems to be a decrease in the carbon content of heat-resisting steels, although at the cost of a reduction in their creep resistance.

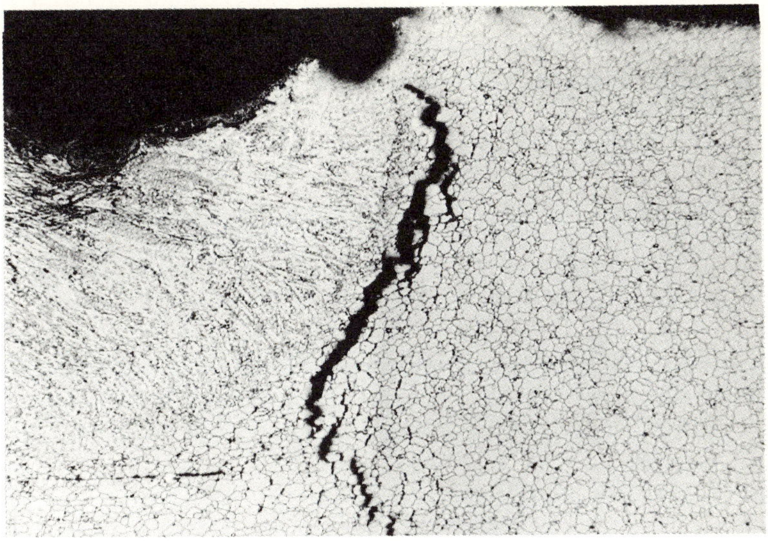

Fig. 6.25. Intergranular fracture in the heat-affected zone of a weld in an 18 % Cr/10 % Ni/Mo/Ti steel (×90).

9 Particular aspects of welding techniques as applied to austenitic stainless and heat-resisting steels

All the fusion welding methods are applicable to austenitic stainless and heat-resisting steels. However, certain special requirements indicated above can have an influence upon the welding procedure.

9.1 Control of the ferrite content

It is relatively easy to control the ferrite content of the weld metal within a limited range when welding with a covered electrode, as the manufacturer can precisely adjust the composition of the covering with respect to that of the core wire. Further, the current to be used is determined by the electrode diameter, and the cooling rate does not vary over a sufficiently wide range to affect the ferrite content of the deposit to any great extent [27]. Special electrodes of this type meant for high or very low temperatures are called 'controlled ferrite' consumables.

The problem is more difficult when welding under argon, because the composition of the weld deposit is almost identical to that of the filler metal. This obliges the steel manufacturer to fabricate stainless steel wires with more restricted compositional ranges than those usually achieved for the fabrication of plates or forgings.

On the other hand, the ferrite content of the weld deposit is reduced as the cooling rate decreases; because of this the high energy processes like spray transfer MIG welding, and especially submerged-arc welding, generally require the use of filler wires having a higher ferritic tendency than the moderate energy processes. The ferrite particles, however, are finer as the cooling rate is higher, or as the heat input is lower[27].

9.2 Limitation of the carbon content

When the working conditions of the welds are such that a risk exists of intercrystalline corrosion, it is essential that the carbon content of the weld deposit be maintained at a relatively low level.

Steel manufacturers now offer 18 % Cr/10 % Ni steels to the user with a carbon content below 0·02 % and this has raised the problem of fabrication of covered electrodes giving a deposit with similar carbon content. It is solved at present by the use of very low carbon wires, and by modification of the coatings so that the carbon contained in the form of carbonates is not transferred in too high a proportion into the deposited metal.

Oxy-acetylene welding of very low carbon grades of steel is not advisable because of the risk of carburisation involved with this process.

Gas-shielded arc welding processes using pure argon, argon–hydrogen mixtures (automatic TIG welding) or argon with 1 % oxygen (MIG welding) do not cause any carburisation of the weld deposit. However, gas mixtures containing a high proportion of carbon dioxide must be prohibited since they lead to an increase in the carbon content of the deposited metal, and can also cause the loss of readily oxidisable elements.

Table 6.6 gives the compositions of weld deposits obtained using the MIG welding process and various shielding gases with an 18 % Cr/8 % Ni/Nb wire[28]. The results show the carbon pick-up and oxidation of niobium caused by carbon dioxide, and illustrate the carbon–oxygen equilibrium which tends to be established during welding. This fixes the carbon content of the weld deposit at about 0·1 % in a carbon dioxide atmosphere.

9.3 Limitation of the heat input

The heat input to the base metal or underlying weld beads must be restricted when conditions are such that there is a risk of intercrystalline corrosion or of sigma phase formation. In this respect, the submerged-arc or spray transfer MIG processes have disadvantages.

Limitation of heat input is also necessary when welding cold-worked austenitic stainless steel, as the width of the recrystallised and thus softened zone depends upon the amount of heat transferred to the base metal. From this point of view, the use of resistance spot or seam welding is advantageous when permitted by the thickness of the workpieces.

9.4 Welding of titanium-stabilised steels

It has already been pointed out in chapter 3, § 4.1, that titanium is almost completely lost into the slag during the fusion of a covered electrode. The

Table 6.6. *Influence of the shielding gas on the composition of weld deposits made using an* 18 % Cr/8 % Ni/Nb *wire*

	C	Si	Mn	Ni	Cr	Nb
Filler wire	0·06	0·62	0·70	8·3	18·6	0·83
Shielding gas						
Argon+1 % O_2	0·06	0·60	0·65	8·3	18·6	0·79
Argon+15 % CO_2+5 % O_2	0·08	0·57	0·60	8·3	18·5	0·83
Argon+20 % CO_2	0·10	0·56	0·62	8·3	18·5	0·79
CO_2	0·13	0·46	0·51	8·3	18·3	0·63

manual metal-arc (or submerged-arc) welding of titanium-stabilised steels for use in the chemical industry is therefore carried out using titanium-free filler metal with a very low carbon content. For high temperature service, a filler metal with molybdenum or tungsten is used (§ 8.2).

However, it is possible to obtain weld deposits containing titanium by the TIG process using pure argon shielding, provided that careful protection of the molten pool from the front and rear sides is ensured. This method is little employed, but can be useful to carry out the root runs of welds for high temperature service, where the internal surface has to resist intergranular corrosion.

9.5 External contamination of welds by copper

Surface contamination of austenitic stainless steel by copper particles, for instance by friction and seizing, can cause cracking in the HAZ during welding under argon. When such surface particles are heated to a temperature between the melting point of copper (1080 °C) and that of the steel, the copper melts without being oxidised, and rapidly penetrates the grain boundaries of the steel, greatly reducing their melting point. This results in intergranular cracking perpendicular to the direction of the welding stress field, the depth of affected metal being greater as the contamination is more severe. Fig. 6.26 illustrates a crack in an HAZ caused by slight surface contamination. Electron probe microanalysis has shown that in the grain boundaries affected by cracking, the local concentration of copper is above 80 %. This phenomenon is specific to the HAZ; in fact, the weld metal can contain several per cent of copper in solution without increasing its cracking tendency.

It is thus important to clean tools employed for working copper very carefully before using them for austenitic stainless steels. When copper backing bars are employed to support and cool the undersides of weld runs, it is essential that the surface condition and cooling arrangement are such that any localised fusion of copper in contact with the steel is avoided; the best precaution consists in the use of nickel plated copper backing bars.

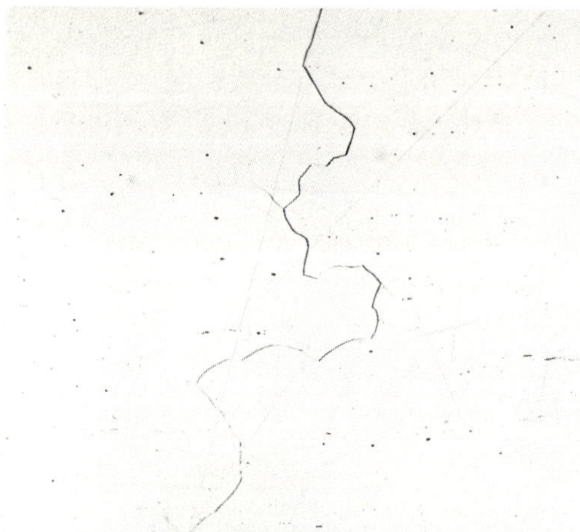

Fig. 6.26. Crack in a heat-affected zone due to intergranular penetration by copper (×90).

References

[1] R. Castro. *C. R. Acad. Sci.* (16 Oct. 1961) 1690.
[2] R. Franks, W. O. Binder and J. Thomson. *Transactions A.S.M.* **47** (1955) 231.
[3] P. Guiraldenq. 5° Journée des Aciers speciaux, Saint-Etienne May 1966.
[4] L. A. Schaeffler. *Metal Progress*, **56** (1949) 680.
[5] D. Séférian. *Métallurgie de la Soudure*, Dunod, Paris, 1959.
[6] W. T. Delong. *Metal Progress*, **77** (1960) 98.
[7] F. C. Hull. *Proc. A. S.T.M.* **60** (1960) 667.
[8] J. Heuschkel. *Weld. J. Res. Sup.* **34** (1955) 484s.
[9] G. W. Oyler, R. A. Matuszeck and C. R. Garr. *Weld. J.* **46** (Dec. 1967) 1006.
[10] B. J. Medovar and Yu. V. Latash. *Avtomaticheskaya Svarka*, **10** (1957) 32.
[11] D. M. Haddrill. *Proc. Conf. Welding creep resistant alloys.* The Welding Institute, London, 1970, p. 126.
[12] L. Pascarel, M. Hubert and C. Messager. *Rev. Met.* **68** (Dec. 1971) 809.
[13] F. C. Hull. *Weld. J.* **46** (Sept. 1967) 399.
[14] A. Randak, W. Wassling, H. Bock, H. Steimauer and L. Faust. *Stahl Eisen* (Oct. 1971) 1255.
[15] W. R. Apblett and W. S. Pellini. *Weld. J. Res. Sup.* **33** (1954) 83s.
[16] P. P. Puzak and H. Rischall. *Weld. J. Res. Sup.* **36** (1957) 57s.
[17] E. Kauhausen and H. A. Vogels. *Metal Progress*, **67** (1955) 129.
[18] N. Cordea, P. A. Kammer and D. C. Martin. *Weld. J. Res. Sup.* **43** (1964) 481s.
[19] L. Colombier and J. Hochmann. *Aciers inoxydables et Aciers réfractaires*, Dunod, Paris, 1965. English translation by Scripta Technica Ltd., *Stainless and Heat Resisting Steels*, Edward Arnold (Publishers) Ltd. London, 1967.
[20] R. A. Daemen and F. Dent, *Weld. J. Res. Sup.* **49** (Jan. 1970) 335–405.
[21] J. Bourrat and J. Hochmann. *Mémoires Scientifiques, Rev. Met.* **59** (1962) 853.
[22] R. Castro, P. Devore and R. Devin. *Mémoires Scientifiques, Rev. Mét.* **59** (1962) 343.
[23] J. Bourrat and J. Hochmann. *Mémoires Scientifiques, Rev. Mét.* **60** (1963) 55.
[24] R. Castro and J. J. de Cadenet. *Rev. Mét.* **62** (1965) 791.
[25] F. A. Hodierne. *J. Iron Steel Inst.* **185** (1957) 225.
[26] R. N. Younger, D. M. Haddrill and R. G. Baker. *J. Iron Steel Inst.* **202** (1963) 693.
[27] T. Tsuk, M. Hubert and C. Messager. *Rev. Met.* **68** (Dec. 1971) 829.
[28] A. A. Smith and G. R. Salter. *Br. Weld. J.* **11** (1964) 222.

7

THE WELDING OF CORROSION-RESISTANT AND HIGH TEMPERATURE NICKEL-BASE ALLOYS

These alloys are a little beyond the scope of this book, because their composition does not allow them to be related to steels. However, many of their applications requiring corrosion resistance or mechanical strength and oxidation resistance at elevated temperature are an extension of the range of application of stainless steels, Moreover, they are sometimes used as filler metals for the welding of steels. For these reasons, it is useful to consider briefly the metallurgical problems arising from the welding of these alloys.

1 General comments – composition and structure

The main nickel-base alloys, apart from commercially pure nickel, are as follows:
– Nickel–chromium alloys, generally containing 15–20 % chromium and up to 10 % iron.
– Nickel–copper alloys, containing 25–30 % copper.
– Nickel–molybdenum alloys, containing 25–35 % molybdenum. There are also a large number of complex alloys containing chromium, molybdenum, tungsten, cobalt, separately or together.

Some of these alloys are employed at around room temperature, under particular conditions when the corrosion resistance of stainless steels is inadequate. This is this case for nickel–copper, nickel–molybdenum and nickel–chromium–molybdenum alloys. With the exception of nickel–copper grades, nickel-base alloys have good chemical and mechanical properties at high temperature and some of the most complex ones are specifically intended for the construction of high efficiency thermal machines, such as gas turbines.

All these alloys also contain minor elements such as carbon, silicon, manganese, and possibly other alloying elements, including titanium, aluminium, niobium and vanadium.

Nickel has a face-centred cubic crystal structure, and is thus an austenite analogous to austenitic stainless steels. All nickel-base alloys have this

Fig. 7.1. Nickel-base alloy with 16 % chromium, 16 % molybdenum, 3·5 % tungsten and 0·65 % silicon, solution treated at 1220 °C and air cooled (×450).

structure, but the numerous alloying and minor elements which they can contain sometimes lead to the formation of complex carbides or intermetallic compounds. Fig. 7.1 shows the microstructure of a nickel-base alloy containing 16 % chromium, 16 % molybdenum, 3·5 % tungsten, 6·5 % iron and 0·65 % silicon, after solution treatment at 1220 °C, and air cooling. It illustrates the high thermal stability of the carbides or intermetallic compounds present in the complex alloys, and also shows the very fast rate of intergranular precipitation which has taken place during cooling.

The precipitation of intermetallic compounds based on nickel, titanium and aluminium is sometimes turned to account to harden the alloy, as in age-hardening austenitic stainless steels (chapter 8, § 2.1).

Apart from the age-hardening grades, nickel-base alloys are usually employed following solution treatment at elevated temperature with rapid cooling. This heat treatment procedure imparts a fully austenitic structure to the simplest alloys, although, as just pointed out, complex alloys can also contain other phases.

2 Microstructural changes on welding

2.1 Simple alloys

In alloys containing few elements forming carbides or intermetallic compounds which are stable at very high temperature (titanium, niobium, molybdenum, silicon etc.), the structure of a weld after cooling is similar to that of a weld in a fully austenitic steel. Fig. 7.2 is an example of such a structure in a nickel–chromium alloy; etching with chromic acid reveals dendritic segregation in the weld metal and intergranular segregation in

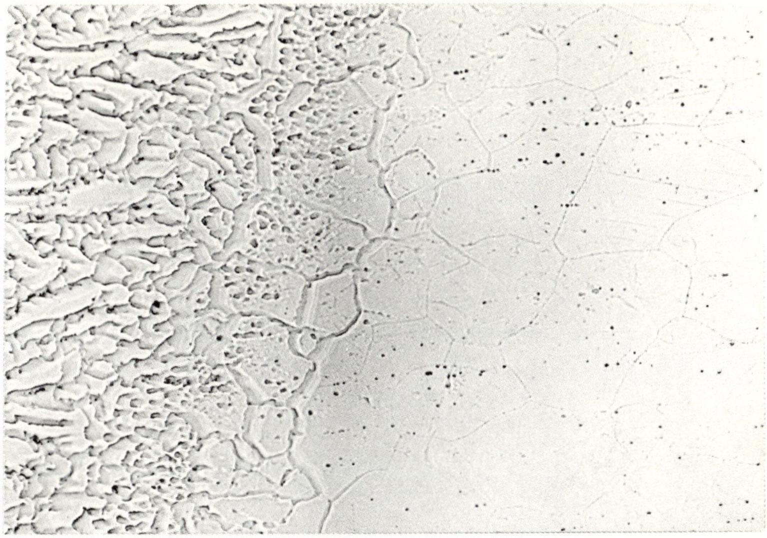

Fig. 7.2. Weld metal and fusion boundary of a weld in a nickel–chromium alloy (×180).

the base metal adjacent to the fusion boundary. In the HAZ of the parent material, the welding thermal cycle has caused fine carbide precipitation.

2.2 Alloys forming carbides or intermetallic compounds which are stable at elevated temperature

In general, the formation of these phases results from the presence of significant amounts of titanium and niobium (carbides), molybdenum (carbides and intermetallic compounds) and silicon (intermetallic compounds). Other elements are also found in the composition of these phases, but have only a slight effect upon their formation.

The presence of these elements causes the appearance of complex eutectics and fine intergranular precipitates in the as-deposited weld metal, as can be seen in fig. 7.3, showing a weld metal in a nickel-base alloy containing 28 % molybdenum, 5 % iron, 0·55 % silicon and 0·02 % carbon. Electron probe microanalysis has shown that the silicon and molybdenum contents of the eutectiform particles are respectively above 2 % and 45 %, while their nickel content is below 48 %. The unaffected parent metal also contains coarse precipitates similar to those visible in fig. 7.1, which can be examined more precisely and easily by microanalysis because of their larger size. It is found that these precipitates contain 62 % molybdenum, 33 % nickel, 4·5 % silicon and about 0·5 % iron. This suggests that one of the constituents of the eutectiform precipitate is similar to the coarse precipitates found in the base metal.

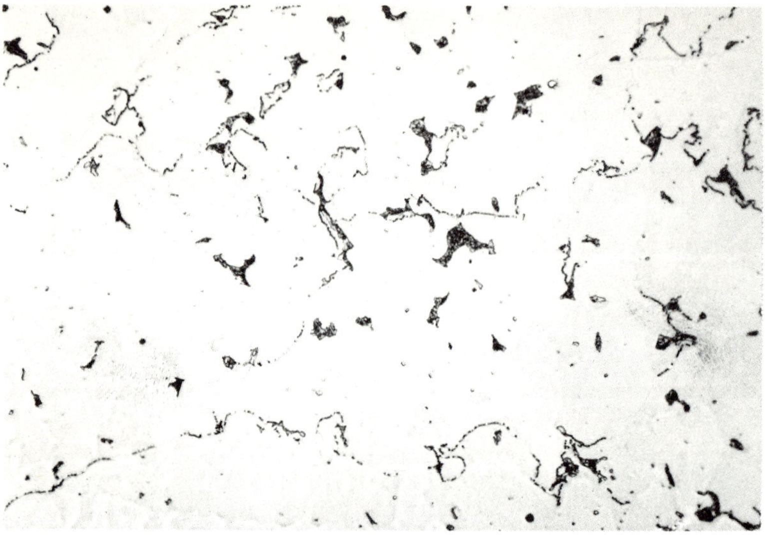

Figure 7.3. Weld metal of a weld in a nickel–molybdenum alloy with 0·55 % silicon (×450).

Table 7.1. *Weld in a nickel-base alloy containing chromium, molybdenum and tungsten*

	Si	Mn	Ni	Cr	Mo	W	Fe
Dendrite axes	0·25	0·5	60	15·8	11·6	5	8·1
Interdendritic regions	0·62	0·7	56	16·4	18·1	4·1	7·5

The formation of this type of constituent in the molten zone results from segregation during solidification. The results of table 7.1 give an example of this segregation in the weld metal of a nickel-base alloy containing chromium, molybdenum and tungsten; they show that during solidification, the remaining liquid becomes enriched in silicon, molybdenum and, to a lesser extent, chromium and manganese, and is impoverished in nickel, iron and tungsten.

In the HAZ, the welding thermal cycle causes two types of microstructural changes. In regions relatively distant from the fusion boundary, fine intergranular precipitation can occur, as illustrated in fig. 7.4, showing the HAZ of a weld in a nickel-base alloy with 22 % chromium, 9 % molybdenum and some tungsten. In alloys containing molybdenum, or chromium and molybdenum, the rate and amount of this precipitation increase with increasing silicon content. This precipitation can be detrimental in certain corrosion situations, or in cold forming operations after welding, when it can lead to intergranular cracking.

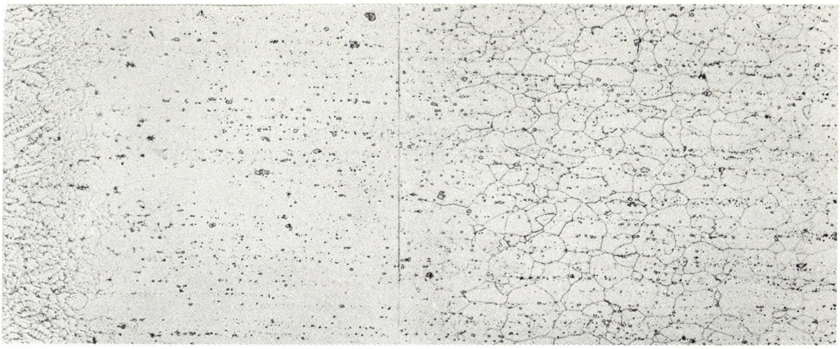

Fig. 7.4. Heat-affected zone of a weld in a nickel alloy with 22 % chromium, 9 % molybdenum and an addition of tungsten (× 75).

In the regions close to the weld metal, the large precipitates and intergranular precipitates existing in the base metal tend to be dissolved, the dissolution being more complete towards the fusion boundary. The rapid dissolution of the large precipitates, which may be either carbides (of M_6C type) or intermetallic compounds, leads to regions of the metal being enriched in the constituent elements, and these zones react differently to chemical attack. This is illustrated by fig. 7.5, in which precipitates can be seen at different stages of dissolution (sometimes complete), surrounded by 'haloes' where the metal is enriched in elements arising from the precipitates. This dissolution sometimes gives rise to a constituent of eutectiform appearance which resembles the precipitates existing in the weld metal. The formation of this constituent has been attributed to an enrichment in titanium of the metal adjacent to the precipitates during their dissolution [1], but it seems that similar phenomena can take place in alloys which do not contain this element.

The dissolution of the intergranular precipitates gives rise to identical phenomena, but on a much smaller scale. When the precipitates are sufficiently dense, however, their dissolution may cause local fusion of the grain boundaries.

3 The formation of porosity in welds

It was pointed out in chapter 3, §1.2, that oxygen and hydrogen (or moisture) can cause the formation of pores in welds in nickel-base alloys, as in stainless steels. With nickel alloys, it is also necessary to consider nitrogen. The solubility of nitrogen in nickel-base alloys diminishes abruptly on solidification, and it has been shown [2] that this gas is the main cause of porosity in nickel welds. As an example, fig. 7.6 illustrates a section through a weld in a nickel–30 % copper alloy made under poor

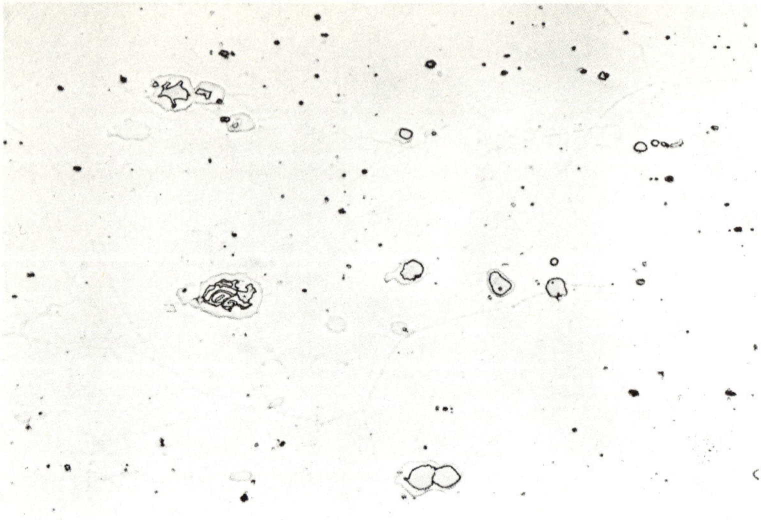

7.5. Dissolution of precipitates close to the fusion boundary (×450).

conditions of shielding against the atmosphere; it will be noted that the root run of the welded joint, which is better protected by the plate edges, contains fewer pores.

To avoid porosity, the joint faces and filler metal should be free from material which is likely to form gaseous decomposition products under the heat of the arc (oxides, greases etc.), while particular care is necessary to protect the molten pool against the atmosphere. These precautions are often insufficient, and a widely employed metallurgical palliative consists in using filler metals containing a little titanium and aluminium. As a result of their high affinity for nitrogen and oxygen, these elements ensure effective degassing of the molten pool. The base metals do not always contain such additions, and it is therefore preferable to weld nickel-base alloys with a filler metal, even when the thickness of the workpiece does not justify it.

In certain cases, the use of argon with small hydrogen additions in autogenous automatic TIG welding enables the number of pores caused by other gases to be diminished, although it is advisable not to exceed 5 % hydrogen to avoid porosity created by the hydrogen itself. It is recommended that only pure argon be used with MIG welding.

4 Cracking at high temperature during welding

4.1 The mechanism of cracking

Nickel-base alloys are sensitive to interdendritic cracking at elevated temperature. The phenomenon as a whole is similar to that encountered in

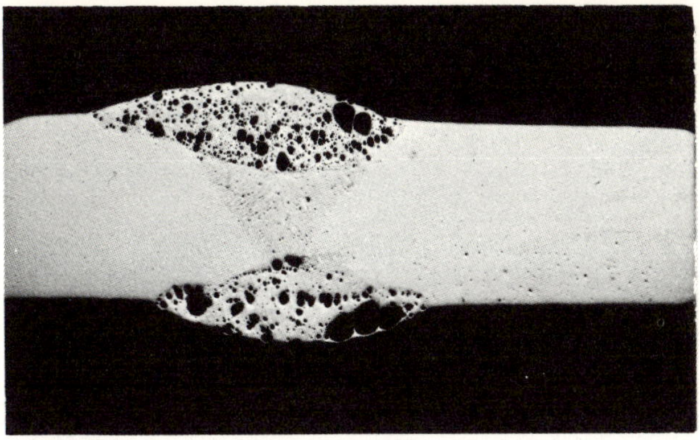

Fig. 7.6. Pores in a weld in a nickel–copper alloy (×2).

stainless steels which are fully austenitic with regard to the Schaeffler diagram (chapter 6, § 4). As with austenitic steels, the cracking can be explained by the presence of liquid or brittle films at the solidification structure boundaries, these films resulting from the segregation on solidification of impurities or minor elements such as sulphur, phosphorus, oxygen, silicon, etc. In nickel-base alloys, it is necessary to include lead in this list. The solubility of lead in nickel alloys is very low, and it is liable to cause cracking by the formation of interdendritic or intergranular liquid films even when it is present only in trace amounts.

Particular attention must also be paid to sulphur, which forms a low melting point eutectic with nickel, and renders the metal very sensitive to cracking. When traces of grease, or of paint or pencil marks containing sulphur are present on the metal surface, this element can even lead to cracking in the HAZ by a mechanism analogous to that described in chapter 6, § 9.5, in connection with the surface contamination of austenitic stainless steels by copper.

In alloys containing elements promoting the formation of carbides or intermetallic compounds, the eutectic which appears in the molten metal can cause interdendritic cracking if its solidification temperature is low enough. Fig. 7.7 is an example of this type of cracking in a weld in a nickel-base alloy containing chromium and molybdenum: it is evident that the cracking follows the eutectiform precipitate, and that it occurred while the precipitate was still liquid. However, it will be seen in §4.3 that the presence of a eutectic is not always detrimental.

Finally, the dissolution of intergranular precipitates in the HAZ, or of the interdendritic precipitates in underlying runs, can reduce the melting temperature of the structure boundaries and cause cracking perpendicular to the stress field in these regions.

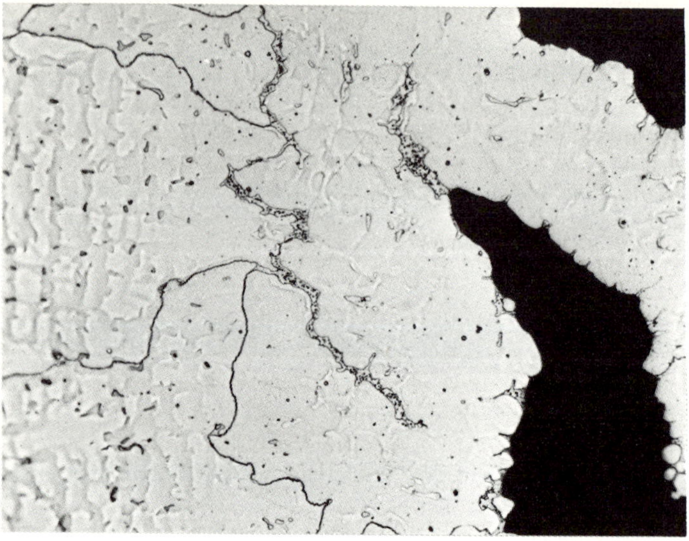

Fig. 7.7. Cracks in the weld metal of a joint in a nickel alloy containing chromium and molybdenum (×450).

4.2 Methods of avoiding cracking

The primary precaution obviously consists of designing the assembly so that the shrinkage stresses are minimal. This is often impossible in alloys having high strength at elevated temperature, since the relaxation of shrinkage stresses is inherently difficult, and these materials are often employed in the construction of highly restrained components for use in gas turbines, for example.

The general rule for welded constructions consists in employing only base metals and especially filler metals with a sufficiently low content of sulphur, phosphorus, oxygen and, when possible, silicon. In particular, the sulphur content of the base metal should not exceed 0·020–0·025 %, and that of the filler metal should be less than 0·015 %.

It is also essential that stringent precautions be taken so that the filler metals and joint edges are not contaminated by traces of grease, paint or pencil marks which are likely to introduce sulphur or lead.

Finally, certain filler metals contain relatively high quantities of manganese (2·5–7 %) and it seems that their resistance to cracking is in consequence slightly improved.

4.3 Particular case – the use of a filler metal with a high niobium content

It was pointed out in chapter 6, § 4.3, that the presence of a liquid phase can be beneficial with respect to cracking at very high temperature, if this phase is present in sufficient quantity to fill the cracks as they are formed until solidification is complete. This principle is used in certain nickel-base filler metals containing 16 % chromium, 10 % iron, some molybdenum and titanium, and 2–3 % niobium. In these alloys, niobium forms a complex eutectic with carbon, nitrogen, silicon and possibly manganese, and it has been found that the presence of about 2 % or more niobium improves the resistance to cracking. When the niobium content is about 1 %, on the other hand, it seems that this element has an adverse effect, as in the case of austenitic steels. Fig. 7.8 shows the microstructure of a weld deposit of this type, containing 2 % niobium.

The ability of niobium-containing eutectics to 'heal' cracks as they form has been demonstrated by welding an austenitic stainless steel containing 17 % chromium, 17 % nickel, 1·8 % manganese and 3·5 % silicon with a nickel-base filler metal having 30 % chromium, 3 % manganese and 3 % niobium. The parent material has a zero ferritic tendency and a high proportion of silicon, and is particularly sensitive to intergranular cracking in the HAZ, similar to that illustrated in fig. 6.10. Fig. 7.9 shows the fusion boundary of this joint, the weld metal being on the left; it is apparent that the base metal contains intergranular cracks, a large proportion of which are filled by intrusions containing a eutectic constituent, surrounded by a wider band more readily etched than the background.

Electron probe microanalysis scans across the whole surface of the sample, give an indication of the local concentrations of the various metallic elements present in these intrusions. The results[3] are summarised by fig. 7.10, showing an optical micrograph of the scanned surface, together with images of the local concentrations of the main alloying elements at the same magnification; on these images, the concentration of the element being considered is higher as the image is lighter. These results confirm that the intrusions originate from the weld metal, accompanied with some diffusion of Mn and Ni on both sides. They further illustrate the part played by niobium and silicon in the formation of the eutectic, complete quantitative analysis of this showing that it contains about 45 % nickel, 26 % niobium, 12 % silicon and only 12 % iron and 3·5 % chromium.

5 Service properties of nickel-alloy welds

5.1 Choice of filler metal

Welded joints in nickel-base alloys are generally homogeneous, i.e. the filler metal, apart from additions of aluminium and titanium, has a composition similar to that of the parent material. Joints made in this manner have properties very similar to those of the base metal. There is generally no metallurgical incompatibility between the various types of alloys, and they can be welded together without difficulty.

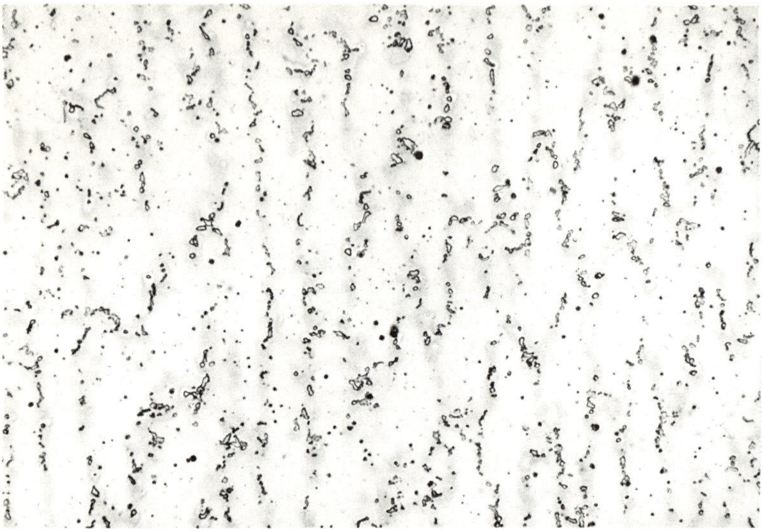

Fig. 7.8. Weld metal of a weld in a nickel–chromium–iron alloy with 2 % niobium (×450).

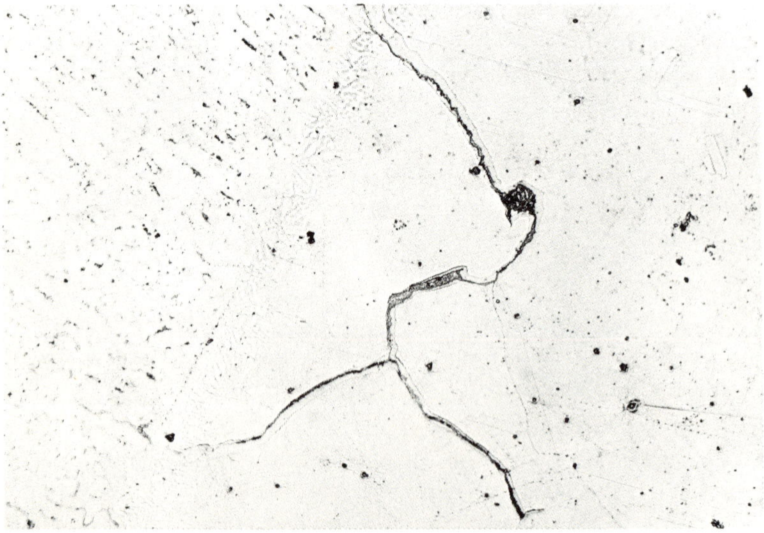

Fig. 7.9. Fusion boundary of a weld in 17 % Cr/17 % Ni/3·5 % Si steel, made using a nickel–chromium filler metal containing 3 % niobium (×180).

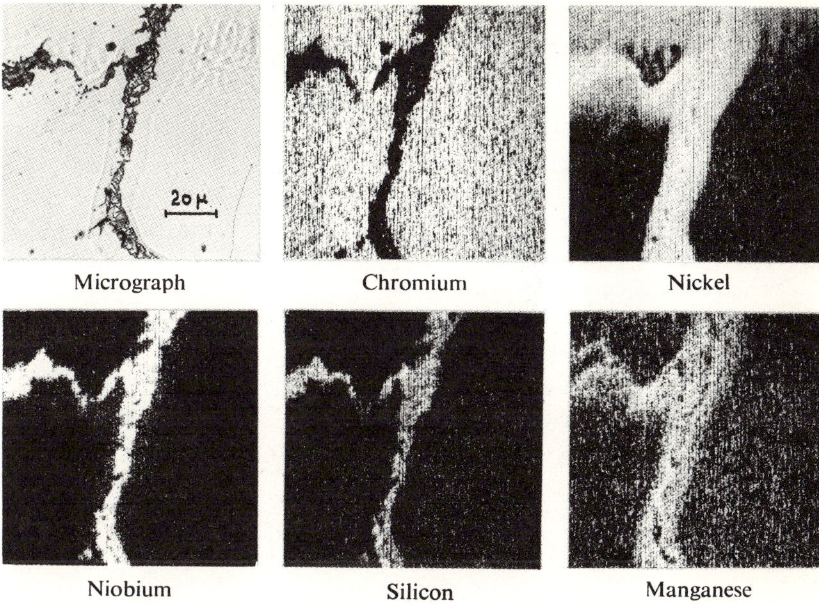

Fig. 7.10. Microanalysis scans of a niobium eutectic intrusion.

5.2 Application of nickel-base consumables for the welding of steels

Nickel-base alloy weld metals, and especially those of the nickel–chromium–iron type, have a number of interesting properties, including the following in particular:

– A high metallurgical tolerance towards dilution by various metals, which enables these alloys to be employed as filler metals for welding all types of steels.

– Good mechanical strength and chemical resistance at elevated temperature.

– Good strength and ductility at room temperature.

– High toughness, even at very low temperature.

These properties are turned to advantage by using the nickel–chromium–iron alloys, possibly with additions of molybdenum and niobium, as filler metals for welds between dissimilar steels, particularly for high temperature service, or for welding less alloyed steels (such as 9 % nickel steel) for cryogenic applications. This type of application will be considered further in chapter 9, dealing with weldments in dissimilar steels.

5.3 Corrosion resistance of welded assemblies

The corrosion resistance at elevated temperature of welds in nickel-base alloys can be reduced when slag from covered electrodes has not been carefully removed. As with stainless steels, welds intended to resist chemical attack must be pickled or ground before service.

It should also be pointed out that, in certain cases, welded joints in nickel-base alloys can suffer localised corrosion having the same effect as intercrystalline corrosion or knife-line attack in austenitic stainless steels.

References

[1] C. P. Sullivan, W. A. Owczarski and W. P. Shirra. *Transactions A.S.M.* **57** (1964) 572.

[2] P. Hancock. *Br. Weld. J.* **11** (1964) 566.

[3] J. Lefèvre, J. J. De Cadenet and R. Castro. 1er Colloque Européen de microanalyse par sonde électronique, Delft (1964).

8

THE WELDING OF AGE-HARDENING STAINLESS AND HEAT-RESISTING STEELS

1 The mechanism of age-hardening

The age-hardening treatment consists of initially retaining certain elements in supersaturated solid solution by quenching, and then treating the alloy in a temperature range where phases rich in these elements are liable to precipitate. The 'pre-precipitation' stage immediately preceding the formation of discrete particles of the precipitate, as well as the stage in which these particles are still very fine and uniformly distributed, can be accompanied by substantial hardening.

The precipitation of sigma phase in austenitic steels is an example of age-hardening. However, this precipitation is accompanied by a marked loss of ductility, and, in general, this type of hardening is not deliberately employed in industrial practice. Hardening mechanisms are required that permit the mechanical strength of the steel to be increased, without excessive reduction of its fracture elongation and toughness.

For practical purposes two types of age-hardening steels can be distinguished:

- Age-hardening stainless steels having an austenitic matrix.
- Stainless steels with a martensite–ferrite matrix, the hardening taking place in the martensite.

Problems arising from the welding of both these categories of steels will be briefly considered, by giving examples.

2 Age-hardening austenitic stainless steels

2.1 General

Austenitic stainless and heat-resisting steels can be hardened during heat treatment by precipitation of phases containing nickel and one or more other elements, such as titanium, aluminium and niobium, when these elements are present in the steel in sufficient quantities.

One of the most widely known steels contains 15 % chromium, 25 % nickel and 2 % titanium, together with additions of aluminium, molybdenum, and vanadium; it will be used as an example.

Following rapid cooling from 950–1000 °C, the hardening elements are either in supersaturated solution in austenite or are fixed in the form of carbides or nitrides, and the alloy possesses the general characteristics of conventional austenites. Precipitation hardening takes place during subsequent heat treatment of several hours at 700 – 750 °C, this treatment giving the steel a yield and tensile strength close to 500 and 900 N/mm^2 (32 and 58 tonf/in.^{2}) respectively, while the elongation and impact value remain at about 25 % and 100 J/100 mm^2.

The improvement of the mechanical properties as compared to conventional stainless steels is retained at high temperature, provided that the service temperature does not reach the range in which the precipitates coalesce ('over-aging'), causing a substantial reduction of the hardening effect; this applies to all age-hardening alloys. In a more general way, it can be said that a relationship exists between the duration and the temperature of the heat treatment, which enables maximum hardening to be obtained; an increase in one of these two parameters without diminution of the other leads to a lower hardening.

2.2 Weldability

From the welding point of view, the 15 % Cr/25 % Ni/2 % Ti austenitic steel taken as an example has the drawback of an entirely austenitic structure with respect to the Schaeffler diagram. Consequently, it is sensitive to interdendritic cracking in the weld metal, as well as to intergranular cracking in the HAZ. The cracks have the same appearance as those observed in the austenitic steels described in chapter 6. It has been found, however, that this type of steel has a higher sensitivity to cracking than non-age-hardenable fully austenitic steels. This is probably due to the fact that the thermal cycle experienced when welding solution-treated material causes precipitation to be initiated in some parts of the HAZ, and of the underlying weld metal in multipass weldments. This preliminary hardening causes significant stiffening of the material, and reduces its capacity for relaxation of shrinkage stresses. This effect is encountered in all types of age-hardening alloys, and results in an apparent increase in the cracking tendency of the weld metal and HAZ. It is also possible that the initiation of intergranular precipitation during the welding thermal cycle leads to the embrittlement of the grain boundaries in the HAZ or underlying runs, reducing their ability to withstand shrinkage strain.

Some physico-chemical causes of cracking can be added to these mechanical effects. Age-hardening steels seem particularly susceptible to the formation of low melting point liquid films by interdendritic segregation, and to the local fusion of grain boundaries in the HAZ at a relatively low temperature. This can lead to the appearance of a crack-promoting interdendritic phase in the weld metal, as shown in fig. 8.1. It can also cause the formation of a grain boundary eutectic in the HAZ close to the

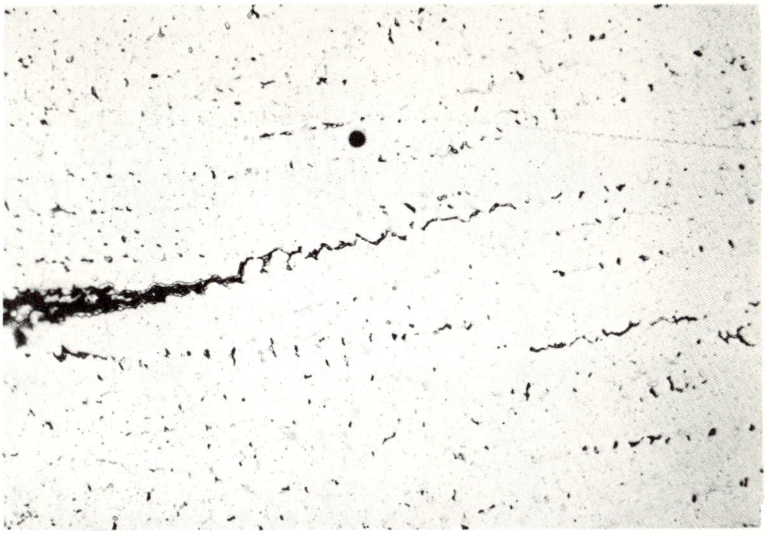

Fig. 8.1. Crack in the weld metal of a weld in a steel containing 15 % chromium, 25 % nickel, 2 % titanium and 0·9 % silicon (×450).

fusion line, as illustrated by fig. 8.2, in which a crack can be seen to have initiated at the weld metal boundary.

It has been found [1] that, in this type of stainless steel, local fusion of the grain boundaries starts at about 1290 °C, and is accompanied by the formation of a eutectic in the liquated grain boundaries. Microanalysis has indicated that these boundaries contain up to 9·7 % titanium, and it is thought that this eutectic is similar to that which appears in the iron–titanium system. Silicon has a high segregation coefficient in austenitic stainless steels (chapter 6, §§ 2.3, 4.2 and 4.4), and may also play a part in the formation both of this eutectic and of the interdendritic constituent in the weld metal. It has, in fact, been shown that a reduction in the silicon content has the effect of decreasing the amount of these constituents in welds, as can be seen by comparing figs. 8.2 (steel with 0·9 % silicon) and 8.3 (steel with 0·2 % silicon). Finally, it has been demonstrated [2] that the presence of phosphorus lowers the melting point of the last traces of liquid still further, down to about 1220 °C by the formation of a phosphorus eutectic. This eutectic may be caused by the coalescence and fusion of titanium phosphide inclusions, and Ni, Ti and Mo silicides. Thus, apart from any benefit obtained by lowering the silicon content, a substantial reduction in the phosphorus content further contributes to improve weldability.

There are also nickel alloys based on the nickel–chromium and nickel–copper systems, which can be age-hardened as a result of additions of

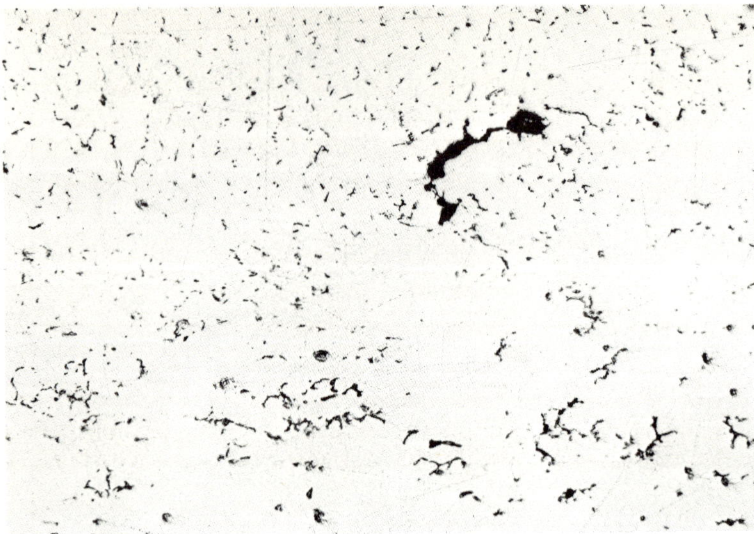

Fig. 8.2. Fusion boundary of a weld in a steel containing 15 % chromium, 25 % nickel, 2 % titanium and 0·9 % silicon (×450).

Fig. 8.3. Fusion boundary of a weld in a steel containing 15 % chromium, 25 % nickel, 2 % titanium and 0·2 % silicon (×450).

titanium and aluminium. The above considerations on cracking during welding are also applicable to these types of alloys, although their susceptibility to grain boundary liquation and to interdendritic segregation and weld metal cracking is less than that of steels containing 15 % chromium and 25 % nickel. Further, age-hardening nickel-base alloys have been developed with improved weldability resulting from attention to the physico-chemical and mechanical causes of cracking. A representative alloy of this type contains 19 % chromium, 18 % iron and 3 % molybdenum, and is age-hardenable by virtue of a niobium content of nearly 5 %, together with small quantities of titanium and aluminium. It was pointed out in chapter 7, § 4.3, that the presence of sufficient niobium improves the resistance of nickel-base alloys to cracking at elevated temperature. Moreover, age-hardening by the precipitation of niobium-rich phases is slow, and a welding operation on solution-treated material causes only negligible hardening in the HAZ.

2.3 Production of welded joints

When welding these steels and alloys, it is important to take all precautions necessary to reduce the intensity of shrinkage stresses. In particular, it is advisable for the workpieces to be in the solution-treated condition, and not age-hardened.

The welding of relatively thin pieces (sheets up to about 3 mm (0·12 in.) in thickness) is generally carried out with a filler metal having the same composition as the base metal. The practical difficulties of the welding operation increase along with the thickness of the workpieces, and it may become impossible to obtain a homogeneous welded joint. This is the case, for instance, with 15 % Cr/25 % Ni/2 % Ti steels. A non-matching filler metal which is not liable to age-hardening is then employed, the composition depending upon the application being considered; it is possible, for example, to use a nickel–chromium alloy with 2 % niobium.

It is apparent that the welding of an age-hardening steel is a delicate operation. Nonetheless, these steels are reliably and consistently welded with matching composition consumables in the fabrication of highly restrained, thin gauge components, such as those in aircraft gas-turbines.

After welding, the joints must be subjected to an age-hardening heat treatment. When a homogeneous welded joint is obtained, the properties of the weld are similar to those of the base metal, although with a slight reduction in the tensile strength and toughness; these differences are reduced when it is possible to carry out solution treatment and quenching after welding and before age-hardening.

3 Age-hardening martensitic stainless steels

3.1 General

These steels are chromium–nickel alloys having an austenite–ferrite structure at elevated temperature, and containing elements promoting age-hardening, such as copper, aluminium, titanium, in isolation or together.

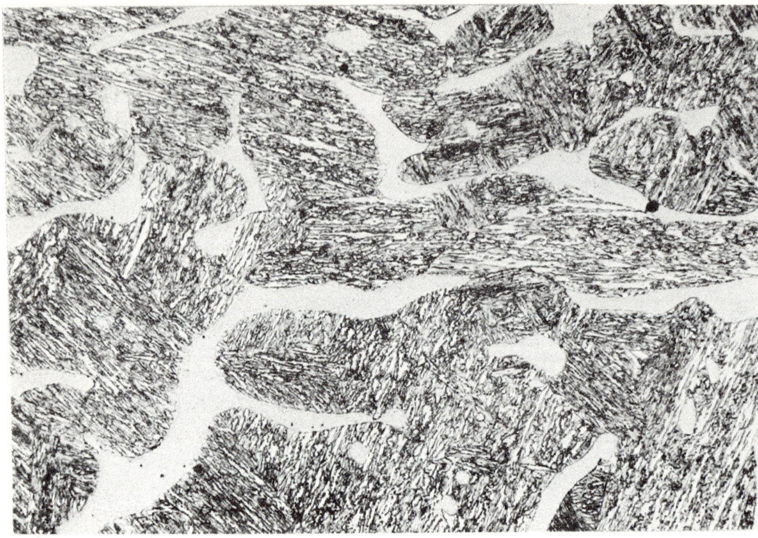

Fig. 8.4. Martensite–ferrite structure of a cast 16 % Cr/4 % Ni/Mo steel containing 2·5 % copper, after age-hardening (×450).

The composition of these materials is adjusted so that the austenite which they contain is unstable, and transforms into martensite which is relatively soft because of the low carbon content. Transformation is induced either directly by quenching, possibly finishing at below room temperature, or by a heat treatment or thermo-mechanical treatment known as 'destabilising' or 'sensitising'. Thus, these steels have a martensite–ferrite structure which can be regarded as lying between 16 % Cr/ 3 % Ni martensitic steels (chapter 4, § 1) and certain austenitic–ferritic steels containing 20 % chromium and 10 % nickel. Fig. 8.4 shows this structure in a cast steel having 16 % chromium, 4 % nickel, 1·5 % molybdenum, 2·5 % copper and 0·04 % carbon, age-hardening being obtained by virtue of the copper addition. The age-hardening treatment consists of a sort of tempering of this martensite during which the usual carbide formation occurs, together with a submicroscopic precipitate. The nature of the precipitating phase has not been well defined, but it results in a considerable hardening of the initially rather soft martensite. According to the temperature and time of the heat treatment, it is possible to obtain a wide range of mechanical properties, and, in particular, to combine a high strength with a satisfactory ductility.

For instance, when quenched from 1100 °C and tempered for four hours at 510 °C, the cast steel mentioned above has a yield and tensile strength of 900 and 1000 N/mm^2 (58 and 63 $tonf/in.^2$) respectively, with an elongation of 12 % and a UF impact value of 70 $J/100\ mm^2$.

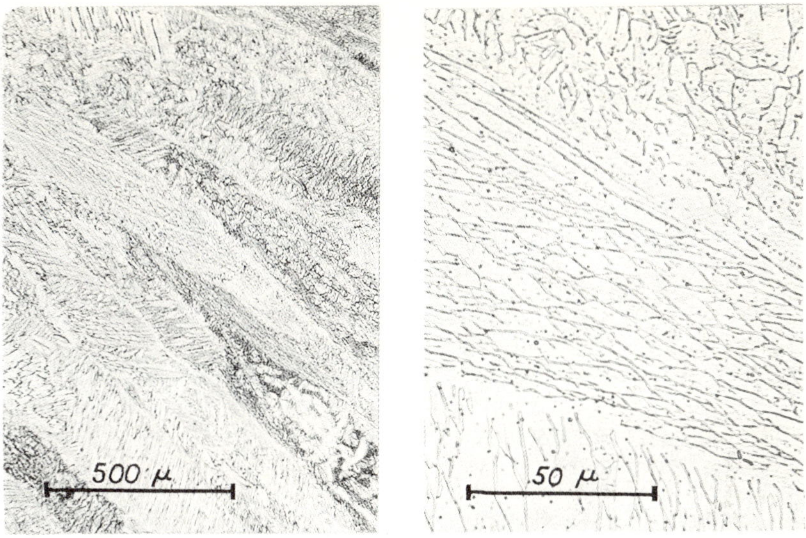

Fig. 8.5. Martensite–ferrite structure of a 16 % Cr/4 % Ni/Mo steel with 2·5 % copper in the as-welded condition.

Other steels of this type have been developed in which the age-hardening results from the presence of aluminium.

3.2 Weldability and weld microstructure

Owing to the austenite–ferrite structure, any risk of cracking at high temperature is avoided. Further, as has been pointed out, these steels have a low or very low carbon content, and the martensite which is formed is relatively ductile. In fact, the hardness of the quenched metal is generally below that which can be obtained by age-hardening. Moreover, the amount of ferrite retained is usually quite high. In consequence, the risk of cracking associated with martensitic steels is greatly reduced, and preheating is not necessary when welding solution-treated and quenched material.

The metallographic structure of welds is rather different from that of the base metal. In the latter, heating to elevated temperature for solution treatment and quenching gives a distribution of phases corresponding to the equilibrium at 1000–1100 °C (fig. 8.4). However, a weld solidifies as delta ferrite, and austenite is formed during the rapid cooling in an analogous manner to that encountered in 17 % chromium semi-ferritic steels (chapter 5, § 2.1). The result is an acicular structure, as illustrated by fig. 8.5, in which the untransformed delta ferrite is present as a very fine network; in some compositions, the as-welded structure can have a certain degree of notch sensitivity.

In this type of alloy, the welding thermal cycle may cause an increase in the amount of delta ferrite present in the HAZ. In rolled products, where

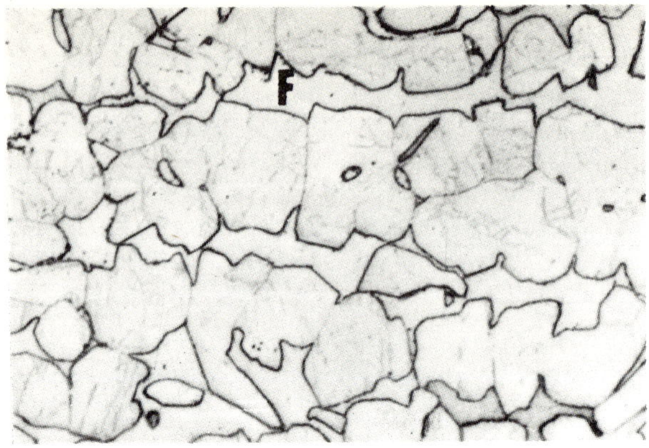

Fig. 8.6. Continuous bands of delta ferrite (F) in the heat-affected zone of an age-hardening stainless steel plate (×900).

the structure is orientated in the direction of rolling, continuous bands of delta ferrite can occur (fig. 8.6), that remain notch sensitive after the age-hardening heat-treatment, and that can cause a brittle fracture.

3.3 Welding and mechanical properties of welds

To reduce the level of shrinkage stresses, it is always preferable to weld components in the solution-treated and quenched condition. The welding of hardened workpieces is possible, as may be necessary for a repair operation, but in this case it is advisable to preheat the joint, particularly if heavy section material is involved. These steels are generally welded with an age-hardenable filler metal with a composition similar to or only slightly different from that of the base metal.

It is possible to obtain joints having practically the same mechanical properties as those of the base metal by subjecting the assembly to a complete solution treatment and age-hardening operation. This method is not often applicable because of the high solution treatment temperatures involved. The most widely used procedure consists of welding hyper-quenched metals, and subjecting the assembly only to an age-hardening treatment, possibly preceded by a destabilising treatment. This method gives satisfactory results, but it does not, however, result in the mechanical uniformity achieved by the complete treatment, as will be seen with an example.

This concerns a 0·07 % carbon steel containing 15 % chromium, 7 % nickel, 2·3 % molybdenum, with 1 % aluminium as a hardening element. In the solution-treated and quenched condition, the structure of this steel consists principally of austenite and ferrite. It is given a destabilising treatment at around 700–750 °C, during which carbide precipitation takes place. The austenite is thus impoverished in alloying elements, and

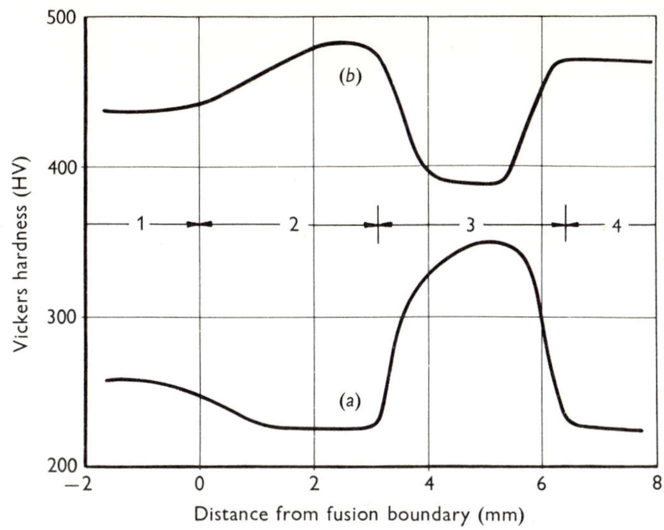

Fig. 8.7. Hardness traverse across a weld in solution treated and quenched 15 % Cr/7 % Ni/2·3 % Mo/1 % Al steel sheet. (*a*) As-welded condition; (*b*) after age-hardening; 1, weld metal; 2 and 3, heat-affected zone; 4, unaffected metal.

transforms into martensite during the subsequent cooling. Age-hardening is then carried out by heat treatment in the range 550–600 °C.

Figure 8.7(*a*) shows a hardness traverse across a section through an argon TIG weld in 2 mm (0·08 in.) thick sheet in the solution-treated and quenched condition. The weld was made with a filler metal having the same composition as the parent material. The hardness of the weld metal 1 is slightly higher than that of the unaffected parent material 4, as a result of its particular microstructure. The HAZ 2 close to the fusion line has been subjected to a high temperature heat treatment similar to a solution treatment, and its hardness is the same as that of the unaffected metal. Between these two zones, there is a region 3, in which the welding thermal cycle has caused destabilisation of the austenite: the structure is composed of ferrite and martensite, thus accounting for the increase in hardness.

Figure 8.7(*b*) represents the hardness curve of the same weld after a destabilising heat treatment at 700 °C, and age-hardening at 600 °C. These two treatments have raised the hardness of the base metal and that of the HAZ adjacent to the fusion line up to practically identical levels, but region 3, the hardest in the as-welded condition, has hardened only very slightly. This is due to the fact that the austenite in region 3 had already been transformed into martensite prior to the destabilising treatment at 700 °C after welding. Heat treatment at 700 °C has resulted in a precipitation which is relatively coarse, having been carried out at too high a

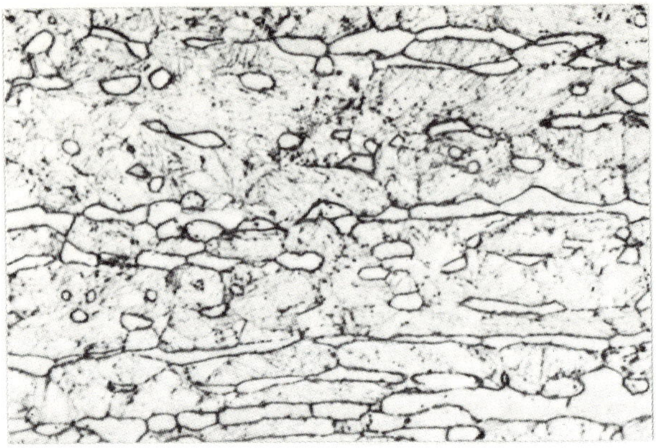

REGION 3

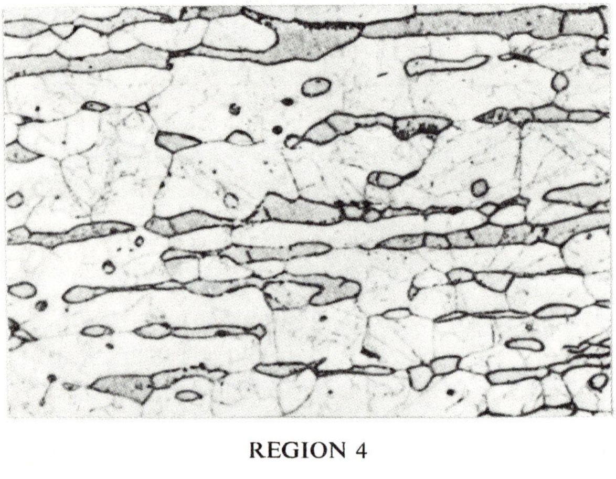

REGION 4

Fig. 8.8. Structures of the regions 3 and 4 of the weld considered in fig. 8.7 after age-hardening (×900).

temperature, and has only a very slight hardening effect (§ 2.1). Fig. 8.8 enables the structure of region 3 to be compared with that of the metal outside the HAZ. This precipitation at elevated temperature suppresses any further possibility of hardening by treatment at a lower temperature.

It may also be noted that the hardness of the weld metal after heat treatment is only slightly inferior to that of the base metal.

This example is an extreme case, as a result of the necessity for a destabilising treatment of the austenite. In grades of steel which do not require this treatment, such as those with a ferrite–martensite structure in the solution-treated and quenched condition, the welding thermal cycle again causes the initiation of precipitation in the martensite at too high a temperature. Hence, a depression in the hardness curve still exists, but it is less marked than in the preceding example.

4 Remarks

4.1

All the welding processes are applicable to age-hardening stainless steels. However, some of these steels contain titanium or aluminium, and this virtually prohibits the manual metal-arc or submerged-arc processes.

4.2

As pointed out, it is advisable to weld these steels and alloys prior to age-hardening. If it is impossible to carry out the hardening treatment on the whole assembly, the advantages of using these materials are largely lost. However, even in the hyperquenched state, certain of these steels have mechanical properties superior to those of conventional stainless steels, and, in some cases therefore, there may be interest in their use in welded constructions without further heat treatment.

References

[1] B. S. Blum and R. H. Witt. *Weld. J. Res. Sup.* **42** (1963) 365 s.
[2] T. Geiger and F. Hengerer. *Rev. Mét.* **64** (March 1967) 263.

9

DISSIMILAR METAL WELDING USING AUSTENITIC FILLER METALS

1 General comments

Dissimilar metal joints involving stainless steels are often encountered in practice. These heterogeneous assemblies can be grouped into three main categories:

(*a*) Assemblies where it is inherent in the construction that welds be made between two steels or alloys either of different types, or having greatly different chemical compositions. The weld metal is then different from at least one of the two base metals.

(*b*) Overlaying of ordinary or low alloy steels using stainless materials in order to protect their surface against corrosion. Hard or wear-resistant overlays may also be mentioned, but these require special alloys which are outside the scope of this book.

(*c*) Assemblies of similar metals in which it is advantageous to use a filler metal of a different nature from the base metal.

The following will be limited to the most frequent practical cases where the filler metal of the dissimilar metal weldment is an austenitic stainless steel or nickel-base alloy.

This situation is encountered in joints of type (*a*) above, when one of the two metals to be welded is an austenitic stainless steel or nickel alloy, or a ferritic stainless steel. Austenitic fillers can also be used in cases (*b*) and (*c*), but they are not always essential. Thus, for instance, a martensitic stainless steel can be welded to an ordinary or low alloy steel by employing a filler metal of a composition similar to that of one or other of the base metals; this type of weld does not involve any particular metallurgical problem, except sometimes in the selection of the heat treatment temperature after welding.

For the overlaying of ordinary or low alloy steel, austenitic stainless steels or nickel alloys are very often selected in view of their ready weldability.

In the third type of heterogeneous weldment (*c*), the use of an austenitic filler metal enables problems associated with martensitic hardening or notch sensitivity of the weld deposit to be avoided in the following cases:
- Welding of air-hardening carbon or low alloy steels.

– Welding of martensitic and ferritic stainless steels.
– Welding the above mentioned steels together.

It will also be seen below that the use of nickel-base filler metals to weld less highly alloyed materials can be advantageous with respect to certain service applications of welded joints.

Two types of metallurgical problems arise in heterogeneous welds with an austenitic filler metal, and will be considered successively:
– The weld metal structure can be wholly or locally changed as a result of the mixing of the filler and base metals during welding. This phenomenon, known as 'dilution', can cause the formation of brittle martensitic structures in the weld metal, or increase its sensitivity to cracking at high temperature.
– The fusion boundary constitutes a discontinuity in chemical composition and physical properties, which can affect the service properties of the assembly. A solution must be found to both these types of problems by selecting a suitable filler metal.

2 Structures of dissimilar metal welds with austenitic stainless steel filler metal

2.1 Dilution

The phenomenon of dilution is illustrated by fig. 9.1, showing the cross-section of a multipass fusion weld made with added filler metal, together with the initial profile of the edge preparation of the joint. The weld metal consists of a mixture of the filler metal and the base material, in amounts respectively proportional to the area of weld metal outside the initial edge profile and to the area between the edge profile and the fusion boundary. The overall dilution is the proportion of base metal which is present in the final weld metal.

This overall dilution has little metallurgical significance, as the weld metal does not have sufficient time to become completely homogeneous, especially when a multipass weld is involved. In practice, the real dilution is higher than the overall dilution in the vicinity of the fusion line, and lower in the regions of weld metal which are furthest from the parent material.

The real dilution increases together with the overall dilution, and is higher as the weld penetration is increased. As an example, the dilution obtained with covered electrodes varies from approximately 25 to 40 %, and can reach 50 % for the root runs; in automatic submerged-arc welding, it can be as high as 70 %. In the MIG process, the dilution is extremely variable according to the current and the selected mode of transfer: it can be very low with the intermittent short-circuit technique, and very

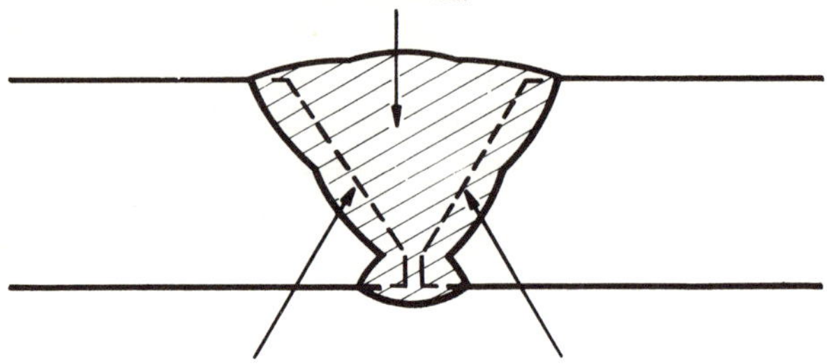

Fig. 9.1. Dilution.

high in the case of spray transfer under high current conditions. The manual TIG process using added filler metal also permits control of dilution: it is relatively low when the operator maintains the arc on the previously deposited metal, but may be high when the arc is held on the base metal. Moreover, for any one process, local variations in dilution can occur when the metal transfer takes place by successive drops [1].

The result of the above is that the choice of a filler metal for a heterogeneous assembly must not be made independently of the welding process under consideration.

2.2 Structures resulting from dilution

For practical purposes, three distinct regions can be identified in a dissimilar metal weld, from the heat-unaffected base metal up to the weld metal furthest from the fusion line:

- The heat-affected base metal.
- A molten zone affected by dilution.
- The weld metal composed of practically pure filler metal.

The microstructures liable to be formed in the weld metal zone affected by dilution can be predicted by means of the Schaeffler diagram shown in fig. 9.2, so long as it can be assumed that the dilution is the same for all the elements involved in the composition of the base material and filler metal. It will be seen below (§3.1) that, in practice, this assumption is justified, at least for metallic elements.

Consider, for instance, the welding of a 9 % nickel steel intended for very low temperature service, which is represented by point *B* in the Schaeffler diagram (fig. 9.2). Homogeneous welds in this steel do not give very good results, as they are often brittle at very low temperature. However, an austenitic filler metal can be employed, containing, for example, 20 % chromium, 13 % nickel, 3 % molybdenum and a ferrite content of about 6 % (see chapter 6, §7); this is represented by point *C*. According to the Schaeffler diagram, the points representing the weld

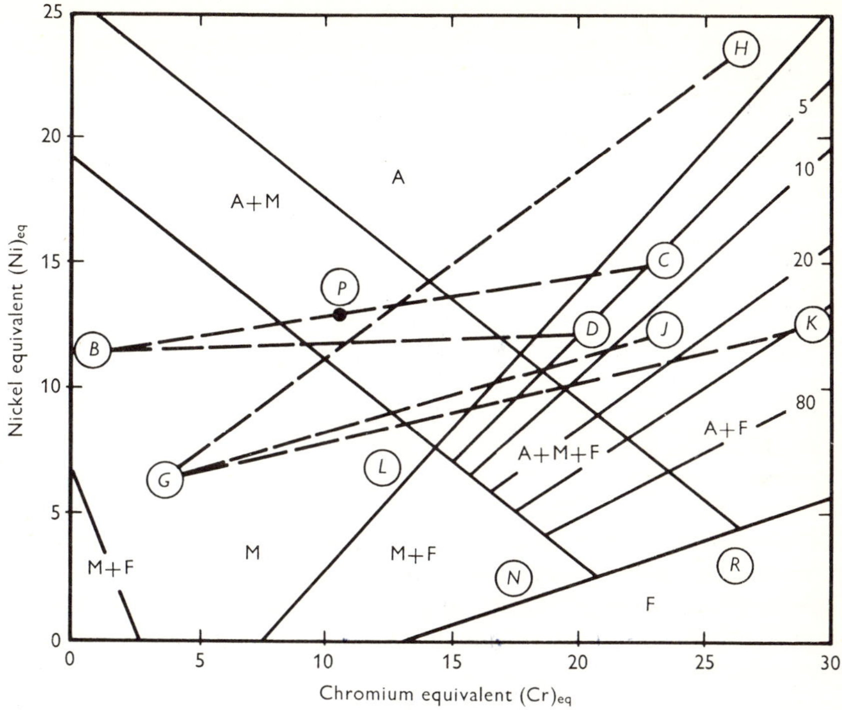

Fig. 9.2. Use of the Schaeffler diagram for dissimilar metal welding problems.

metal structures obtained by mixing *C* and *B* lie on the straight line *CB*; moreover, the proportion of *B* which must be incorporated in the weld metal (i.e. the dilution) to obtain the structure of any point *P* on this straight line, is equal to *CP*/*CB*.

In this example, the structures that can be encountered from the undiluted filler metal up to the fusion line are successively:

- Austenite–ferrite with less and less ferrite.
- Pure austenite.
- Austenite–martensite at a dilution equal to 37 %.
- Pure martensite at a dilution equal to 65 %.

These structures are, in fact, encountered in practice, and fig. 9.3 shows the boundary between the fully austenitic and austenite–martensite regions in a weld in 9 % nickel steel made using a 20 % Cr/13 % Ni/3 % Mo filler metal.

It can be noted in this example that, despite the choice of a filler metal with a sufficiently high ferrite content for a homogeneous weld to be insensitive to cracking at high temperature, the dilution causes the formation of a fully austenitic region which can create difficulties during the welding operation. Generally, this problem can be reduced by increasing

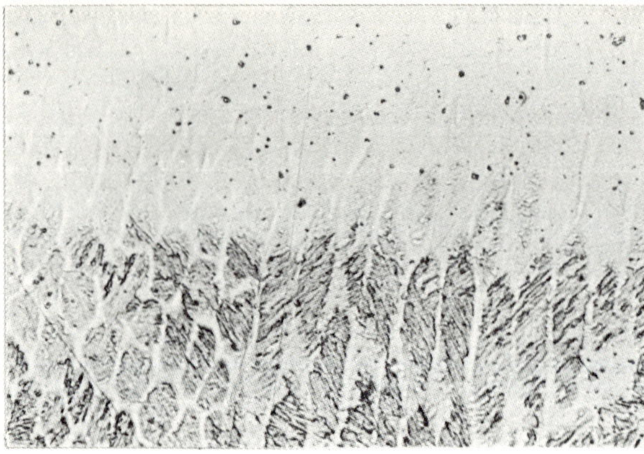

Fig. 9.3. Boundary between the fully austenitic and austenite–martensite zones in a 9 % nickel steel weld, made with a 20 % Cr/13 % Ni/3 % Mo filler metal (× 500). A, austenite; M, martensite.

the chromium equivalent of the filler metal without increasing the nickel equivalent, or at least by increasing it to a lesser extent. This results in an increase in the dilution at which pure austenite is formed; all other things remaining the same, the proportion of weld metal having this structure is reduced. In the present case, this solution is not applicable, as a very high ferrite content would impair the toughness at very low temperature. It is therefore necessary to shift the point representing the filler metal along a line corresponding to a constant ferrite content of about 6 %. If it is shifted towards the upper right corner of the diagram, the dilution at which pure austenite is formed is practically unchanged. It remains the same if it is shifted to point *D*, although the size of the fully austenitic zone is reduced. However, in the case of point *D* the austenite–martensite and pure martensite appear at dilutions of 20 % and 55 % respectively, distinctly below what they were with point *C*. These dilutions of 20 % and 55 % are encountered in practice, and the weld metal thus contains some martensite which may be detrimental to the toughness of the weld. In fact, it is found that the composition represented by point *C* is quite a good compromise.

It should be appreciated that the indications of behaviour given by the Schaeffler diagram must not be considered as rigorous. Nonetheless, as illustrated by this example, the diagram permits sufficiently accurate prediction of the structures resulting from dilution which appear in heterogeneous welds, and it greatly helps in the selection of filler metals.

On the other hand, it should be noted that the points plotted on figs. 9.2 and 9.5 (§2.5) are representative of the composition and structure of weld deposited metal only for compositions similar to those used to establish the diagram, i.e. compositions of conventional grades of stainless steel with relatively high contents of alloying elements. For different compositions, such as high manganese stainless steels, or low alloy steels, the point plotted on the diagram is, of course, always representative of the weld metal composition, but it may not reliably indicate its structure.

Thus, the structures of weld metals with compositions corresponding to points *G* and *B* of figs. 9.2 and 9.5 may not be composed only of martensite or martensite and ferrite respectively. This limitation applies only to the vicinity of points *G* and *B*, and in no way reduces the significance of the straight lines joining them to the points representing the weld metals over the usual range of dilution.

2.3 Welding of air-hardening low alloy steels with an austenitic stainless filler metal

The concept of employing an austenitic stainless filler metal for the welding of air-hardening low alloy steels (steels for armour-plate, chromium–molybdenum steels for high temperature service etc.) originated at the (Paris) Institute of Welding[2] as early as 1938. The initial idea was to 'butter' the edge preparation of an air-hardening steel with an austenitic filler metal, to heat-treat the workpieces thus obtained to temper the martensitic HAZ, and then to join them with the same filler metal. Provided that the thickness of the buttering layer was sufficient, this allowed the final welding to be carried out without preheating or post-weld heat treatment. As experience showed that the use of austenitic electrodes enabled HAZ cracking caused by martensite formation to be avoided, this idea has been developed, and these steels are now welded with the aid of austenitic filler metals without previous buttering, without preheating and sometimes without further heat treatment. This possibility was pointed out in chapter 4, § 4.3, with reference to martensitic stainless steels, and its success can be explained on the basis of the cracking mechanism described in § 3 of that chapter. It will be recalled that weld metal cracking can easily be avoided provided that the filler metal has a sufficient ferrite content and a ductility high enough to bear shrinkage stresses without failure, and that cracking in the HAZ does not occur for the following two reasons:

- Hydrogen remains in solution in the weld metal and does not diffuse into the base metal.
- The high plasticity of the weld metal at elevated temperatures permits the relaxation of stresses in the HAZ, particularly when this zone is transformed into martensite.

The necessity to carry out a heat treatment after welding to soften the martensitic zone should be considered only in relation to the nature of the parent material, and to the service application of the weldment. There are cases in which this region, although brittle, is not considered wide enough to impair the service behaviour of the assembly.

In the welding of air-hardening steels (such as the chromium–molybdenum steel, represented by point *G* on the Schaeffler diagram), the two problems illustrated by the example of §2.2 are found again:
- To ensure good weldability, requiring the selection of a filler metal with a sufficient ferrite content, taking into account the anticipated dilution.
- To avoid the formation of a marked quantity of martensite in the weld metal, leading to the use of a highly alloyed filler metal.

Depending on which of these two aspects is given priority, different solutions to these problems can be adopted, as considered and exemplified more fully in references [1,3,4]. In the U.S.A. and elsewhere, it seems that to avoid the formation of martensite is considered the more important aspect, and this leads to the use of 25 % Cr/20 % Ni type filler metals, despite the risk of high temperature cracking in the weld metal. These filler metals can be represented by point *H* on fig. 9.2, where consideration of the straight line *GH* shows that a fully martensitic structure appears only above a dilution of about 70 %, and this is seldom encountered in practice.

In France, it is generally considered more advisable to employ a filler metal with a sufficiently high ferrite content, usually a 20 % Cr/10 % Ni type with 3 % molybdenum (point *J* of fig. 9.2, for instance). This virtually eliminates all risk of cracking at high temperature, but permits the formation of small amounts of martensite, and especially of an austenite–martensite zone in the most diluted regions. This border of martensite is, however, unlikely to be really dangerous when its size is relatively small. The use of a 24 % Cr/12 % Ni filler metal is similar to this approach, provided that it has a low carbon content.

Intermediate solutions can be conceived, with the aim of taking advantage of the two previous approaches. For instance, a 25 % Cr/20 % Ni filler metal can be employed in the most diluted regions, with the weld being completed using a conventional 18 % Cr/10 % Ni consumable. It is also possible to use a filler metal containing 29 % chromium, 9 % nickel and 0·1 % carbon, which possesses the double advantage of being highly alloyed and having a ferrite content close to 40 % (point *K* of fig. 9.2). Further, as seen in chapter 6, §6.1, 29 % Cr/9 % Ni type deposits have a high mechanical strength and a reasonable toughness, and this permits their use for the welding of armour-steels for example; unfortunately, the high ferrite content of the welds prevents their application in assemblies meant for high temperature service, because of the increased risk of embrittlement by sigma phase formation.

2.4 Welding of martensitic and ferritic stainless steels with austenitic stainless filler metals

In chapter 4, §4.3, and chapter 5, §4.4, the possibility was pointed out of welding martensitic and ferritic stainless steels with austenitic stainless filler metals. The advantages of this method result, as in the preceding case, from the absence of sensitivity to weld metal cracking, from the fact that hydrogen diffusion to the base metal is avoided, and from the high ductility of the weld deposit.

For instance, suppose that the composition of a martensitic steel is represented by point *L* of fig. 9.2. If this metal is welded using an 18 % Cr/10 % Ni or 18 % Cr/10 % Ni/Mo type filler metal, the composition of which is represented by a point near *C*, *D* or *J*, the Schaeffler diagram shows that the structures resulting from dilution cannot contain pure austenite, in contrast to the previous case. In fact, it is found that the structures encountered consist of austenite, martensite and ferrite, or austenite and ferrite. The use of a 25 % Cr/20 % Ni consumable gives rise to structures resulting from dilution similar to those observed in low alloy air-hardening steels.

The necessity for heat treatment after welding should be considered only as a function of the HAZ structure.

In the same way, if the filler metals *C*, *D* or *J* on fig. 9.2 are used to weld a 17 % chromium semi-ferritic steel represented by point *N*, the structures resulting from dilution will be of the martensite–ferrite–austenite and austenite–ferrite types. An example of these structures is given in fig. 9.4 which shows the fusion boundary of a 17 % chromium steel weld made with a filler metal having a very low carbon content and a ferritic tendency of 8 %. In this particular case, the high dilution austenite–ferrite–martensite zone is very thin. It is also apparent in this figure that the austenite–ferrite region of the weld metal adjacent to the fusion line contains a distinctly lower proportion of ferrite than that predicted by the Schaeffler diagram. Use of the straight line *NC* in fig. 9.2 indicates that the diluted weld metal should in fact have a ferrite content above that corresponding to the pure filler metal (8 %), contrary to fig. 9.4. This is quite frequent when a very low carbon austenitic filler metal is employed to weld a steel with an average or high carbon content, and it may result from the fact that the diffusion coefficient of carbon, an austenite former, is higher than that of other metallic elements.

Finally, the welding of 25 % chromium ferritic steels, represented by point *R* in fig. 9.2, can be carried out with a 25 % Cr/20 % Ni filler metal (point *H*) because, in this case, only a low dilution is sufficient to cause the formation of an austenite–ferrite structure in the weld metal; these refractory steels are also welded with 29 % Cr/9 % Ni type filler metals (point *K*).

2.5 Overlaying of mild steel with an austenitic stainless steel filler metal

This type of overlaying is performed in order to protect the mild steel surface against corrosion. Filler metals of a martensitic stainless steel or giving a deposit with a ferrite–martensite structure are sometimes

Fig. 9.4. Dilution zone of a weld in 17 % chromium steel made with an 18 % Cr/10 % Ni filler metal having a ferritic tendency of 8 % (×500).

employed, but most overlaying is carried out with 18 % Cr/10 % Ni or 18 % Cr/10 % Ni/Mo filler metals, as these grades can be deposited very easily.

For example, consider a base metal of composition represented by point *B* on fig. 9.5 being overlaid using a 17 % Cr/9 % Ni/3 % Mo type filler metal represented by point *C*. The Schaeffler diagram shows that, as in welding martensitic stainless steels, the structures resulting from dilution which are encountered in going from the base metal to the slightly diluted weld metal must be the following:

- Pure martensite.
- Martensite–austenite–ferrite, possibly with a small zone without ferrite.
- Austenite–ferrite.

This is found in practice, as shown by fig. 9.6 which represents the fusion boundary of this type of weldment.

In this case, the presence of a martensitic region is rarely detrimental from the point of view of mechanical properties. However, a more highly alloyed metal can sometimes be employed for the first layer in order to increase the chromium content of the highly diluted layer, and obtain a greater thickness of stainless metal.

To obtain the same result, it is also possible to select the welding process or control the welding conditions to reduce the penetration of the first deposited layer. Thus, it is preferable to operate covered electrodes at quite a low current, or to employ MIG welding with globular transfer, and

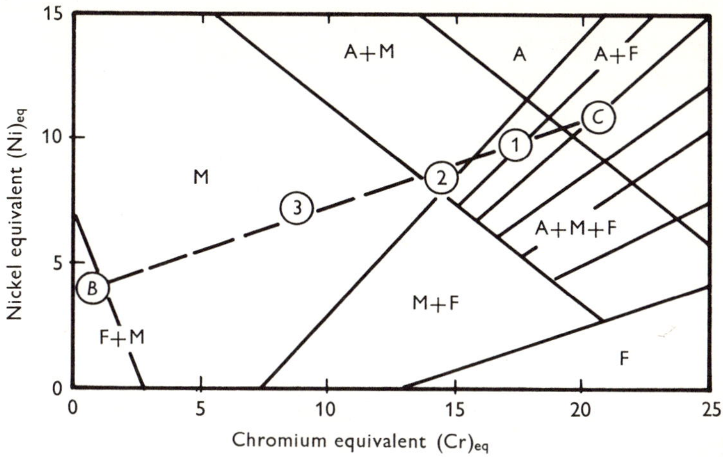

Fig. 9.5. Overlaying a mild steel with a 17 % Cr/9 % Ni/3 % Mo consumable. A, austenite; F, ferrite; M, martensite.

direct polarity if necessary (chapter 3, §6.3). The submerged-arc welding, in spite of its high deposition rate, offers little advantage for the first overlaying run if a filler metal with a high chromium content is not used.

The structures and applied properties of these overlayings will be considered further in §3.1.

3 Applied properties of dissimilar metal weldments made with an austenitic stainless steel filler metal

The applied properties of these assemblies result, in the first instance, from the properties of the individual component parts. Consequently, the effect of welding is the same as in the cases of homogeneous welding studied in the previous chapters. However, the presence of a metallurgical bond between two metals having different physical and chemical properties can create particular problems, two examples of which will be given in this section.

3.1 Corrosion resistance of heterogeneous assemblies – stainless steel overlays

The corrosion resistance of a dissimilar metal weldment is, of course, limited by that of the less resistant component of the assembly. However, some anomalies in corrosion behaviour can occur because of the formation of galvanic cells resulting from the differences in composition which exist between the parts of the weld. Effects arising from dilution can also be detrimental to the corrosion resistance.

For instance, when an ordinary steel component is welded to a stainless

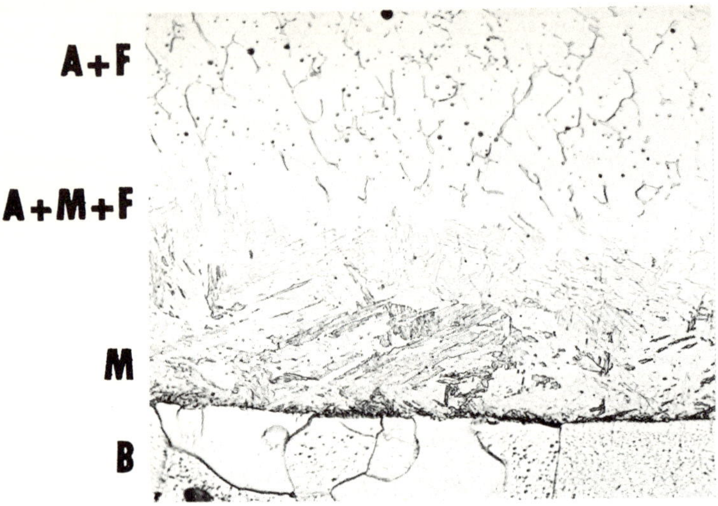

Fig. 9.6. Fusion boundary of an overlay on mild steel using a 17 % Cr/9 % Ni/3 % Mo filler metal (× 500). B, base metal; M, martensite; F, ferrite.

steel pipe or vessel, carbon from the plain steel may diffuse into the stainless steel during welding, or in service at elevated temperature (§3.2). This can impair its corrosion resistance if the assembly is in contact with a corrosive fluid. In this case, it is advisable first to weld a plate of stainless steel onto the vessel or piping, and then to weld the ordinary steel to this plate, in order to remove the dissimilar metal weld from the dangerous zone.

In the same context, post-weld heat treatment of a martensitic stainless steel weld made with an austenitic stainless steel filler metal can cause the sensitisation to intercrystalline corrosion of the weld metal regions enriched in carbon by dilution.

When low chromium steels are welded with a stainless steel, the dilution can affect the corrosion resistance by lowering the chromium content of the deposit and taking it below the 'stainless threshold'. This problem must be taken into consideration when an ordinary steel is weld-overlaid with a stainless steel, and when the cladding is liable to be thinned by wear or corrosion. It is thus important to know the thickness of the cladding in which the metal is stainless, so that a repair can be carried out before the underlying metal is exposed which would then be quickly destroyed by the corrosive medium.

As an example, fig. 9.7 shows an etched metallographic section of an overlay containing 17 % chromium, 9 % nickel, 3 % molybdenum and 0·04 % carbon, deposited onto a mild steel with 0·15 % carbon. The base metal is strongly etched, and appears dark in the lower part of the figure, while the slightly diluted metal at the top of the overlay is little attacked;

(a)

(b)

(c)

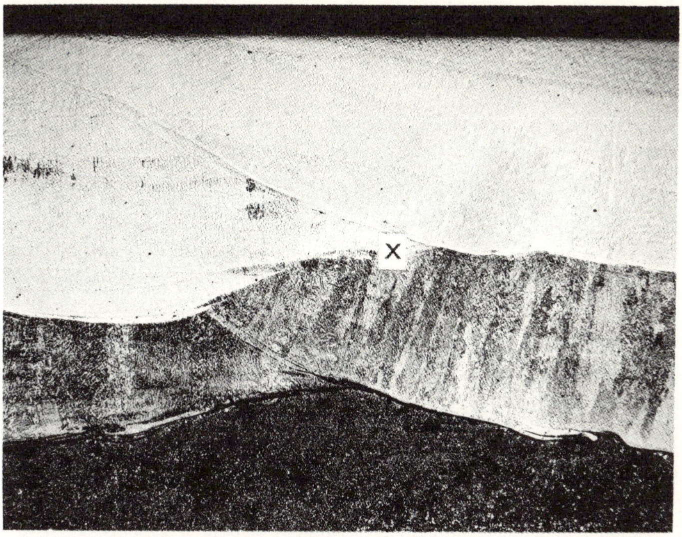

Fig. 9.7. Overlay on mild steel deposited using a 17 % Cr/9 % Ni/3 % Mo filler metal (× 15). (*a*) Slightly diluted metal; (*b*) highly diluted zone; (*c*) base mild steel.

between these two regions there is a highly diluted zone containing a marked proportion of martensite. Electron probe microanalysis has shown that the first martensite laths encountered in going from the slightly diluted weld metal towards the base metal (i.e. in the direction of the decreasing chromium), have a chromium content of between 13 and 14 %. The formation of these laths which are clearly visible on a metallographic section, can consequently be taken as a reference to estimate the overlaying thickness in which 'stainlessness' is assured; this limit is marked by a cross on fig. 9.7.

The microanalysis determinations carried out with an electron probe on such an assembly also permit verification of the assumption made in §2.2, that structures likely to be formed by dilution are indicated on the Schaeffler diagram by the straight line joining the points representing the base metal and filler metal compositions. In the above example, these points correspond to the points *B* and *C* in fig. 9.5. Measurements of composition have been carried out using linear scans, which allow the average composition of the various metal regions to be estimated independently of differences existing between the interdendritic areas and the dendrite axes. Carbon cannot be analysed in this way, and it has been assumed that the proportion of this element varies in a similar manner to the nickel content. The results are given in table 9.1, showing the nickel and chromium equivalents calculated from these measurements, together with the metallurgical structures of the corresponding regions.

Table 9.1. *Chemical composition of different regions of a stainless steel overlay on a mild steel*

Si	Mn	Ni	Cr	Mo	C†	$(Ni)_{eq}$	$(Cr)_{eq}$	Structure	Point
0·6	0·6	9	17	3	0·04	10·5	20·9	Austenite Ferrite	*C*
0·4	0·6	7·5	14	2·7	0·06	9·6	17·3	Austenite Ferrite Martensite	1
0·4	0·5	7	11·5	2·3	0·065	8·2	14·4	Austenite Martensite	2
0·2	0·5	3·8	7	1·5	0·10	7	8·8	Martensite	3

† Calculated values, except for point *C*.

The points corresponding to the analyses in table 9.1 are plotted on the Schaeffler diagram in fig. 9.5, and can be seen to be appropriately situated. Thus, the assumption of straight 'dilution lines' made in §2.2 is justified sufficiently well to be used as a guide in dissimilar metal welding, provided that the carbon dilution follows a law similar to that of the other elements, which is perhaps not always strictly valid because of the possibility of rapid diffusion.

This assumption has also been verified recently, by employing an analogous technique [5] for the examination of the interfaces between ordinary steel and different types of austenitic stainless steel weld metal.

3.2 Dissimilar metal weldments for high temperature service

Austenitic stainless steels may be used as filler metals for welding low alloy or stainless martensitic steels in heterogeneous assemblies intended for high temperature duties. Apart from the risk of embrittlement by sigma phase formation in the weld deposit, particular problems can result from the following factors:

– The base metal generally has a higher carbon content than that of the austenitic weld metal.

– The austenitic stainless steel weld metal has a higher coefficient of expansion and a lower thermal conductivity than those of the parent material.

The differences in carbon content give rise to diffusion of this element when the assembly is maintained at between 550 and 900 °C approximately. Under these conditions, it is always found that carbon diffuses from the low alloy steel or 13 % chromium stainless steel base metal towards the austenitic stainless steel weld metal. This results from the fact that the carbon content of the base metal (about 0·2 %) is generally superior to that of the weld deposit (below or equal to 0·1 %), and in particular from the differences in chemical activity of carbon in the two metals. In effect, the high contents of carbide formers like chromium, and possibly molybdenum, tungsten, titanium and niobium tend to lower the chemical activity of carbon in the weld metal, and favour diffusion of this element towards the weld deposit.

Tests performed on diffusion couples held at elevated temperature [6] have shown that iron, nickel and chromium can also diffuse across the interface between an ordinary steel and an 18 % Cr/10 % Ni steel. Iron diffuses from the ferritic steel towards the austenitic steel, nickel from the austenitic steel towards the ferritic steel, and chromium in both directions simultaneously, at rates depending upon the diffusion of the other elements. Similar diffusion may probably occur in the transition zones of welds, but the diffusion of carbon is without doubt the most important phenomenon in practice.

This carbon diffusion can take place either during heat treatment to temper martensite present in the HAZ, or when the assembly is put into service. It has two main consequences: First, decarburisation and grain growth of the HAZ of the base metal occur with associated reduction in its mechanical properties. The regions close to the fusion line can even become fully ferritic. Second, it leads to carburisation of the weld metal zone adjacent to the fusion line. In 18 % Cr/10 % Ni steels stabilised with titanium or niobium, this carburisation causes precipitation of complex carbides. While carbide precipitation also takes place in unstabilised steels, the formation of martensite can sometimes be seen in addition. This can be explained either by an impoverishment in chromium due to the carbide precipitation, or by the impoverishment in nickel resulting from the diffusion mentioned above. In both cases, the carburisation is manifest by a hardening of the weld metal adjacent to the fusion boundary.

The difference between the expansion coefficients of the base metal and weld metal causes higher internal stresses along the fusion boundary than those existing in homogeneous welds. Heat treatment after welding, when carried out, does not permit their complete elimination; even if these stresses were totally relieved at the heat treatment temperature, further stresses would be set up by cooling after heat treatment.

Differences in expansion coefficient and thermal conductivity are also a source of internal stresses when the service use of the assembly involves thermal cycling. Thermal fatigue will occur, and may sometimes lead to rupture of the joint along the fusion boundary.

The difficulties related to carbon diffusion and to the presence of high stresses resulting from the differences in physical properties can be eliminated by using a nickel-base filler metal of the nickel–chromium–iron type [7]. This type of weld deposit has a coefficient of expansion between that of low alloy or 13 % chromium martensitic steels and that of austenitic stainless steels. Moreover, the chemical activity of carbon is higher in these alloys than in 18 % Cr/10 % Ni and 25 % Cr/20 % Ni steels, which significantly retards the diffusion of carbon across the fusion boundary.

Consequently, it is preferable to use this type of filler metal rather than an austenitic steel, when the dissimilar metal joint is intended to work at elevated temperature under severe conditions of thermal cycling and high stresses.

4 Use of a nickel-base filler metal in dissimilar metal weldments

In the previous paragraph it was pointed out that a high nickel filler metal could be used to weld martensitic steels meant for high temperature service, with certain practical advantages. More generally, nickel-base filler metals can be employed for dissimilar metal joints in a large number of alloys. Pure nickel is sometimes used, but it is more common to select alloys containing 16 % chromium and 10 % iron with additions intended to improve their resistance to cracking at elevated temperature (2 % niobium) and to increase their mechanical strength (molybdenum, tungsten, titanium etc.). This type of weld deposit has the following advantages:

– The dilution of these alloys by high proportions of iron, nickel, chromium etc. does not lead to the formation of structures which are brittle or particularly sensitive to cracking at high temperature. When martensitic steels are welded, the dilutions necessary to cause the formation of significant amounts of martensite in the most diluted weld metal are practically never reached with the usual welding procedures. A martensitic layer of only a few microns thickness may be formed [5], but this is of no practical importance.

– They resist corrosion at room temperature and oxidation at high temperature.

– They have relatively high mechanical strength and ductility at room temperature, and good creep resistance at high temperature.

– They remain very ductile at very low temperature. In particular, their toughness at the temperature of liquid nitrogen is virtually the same as at room temperature.

– They are not sensitive to embrittlement at high temperature by sigma phase formation.

– As already mentioned, their coefficient of expansion is between those of ferritic and martensitic steels and that of austenitic stainless steels.

The combination of these properties permits their use in the following cases:

– Welding different types of nickel-base alloys† together, or to all types

† This includes nickel–copper alloys; however, it may be better to interpose a pure nickel layer between copper-rich alloys and high chromium steels.

Fig. 9.8. Fusion boundary of a weld in 9 % nickel steel (*a*); with a nickel–chromium–iron filler metal containing 2 % niobium (*b*) (×450).

of ordinary or stainless steels, apart from free-machining steels with high sulphur contents. Dilution of nickel-rich weld metals by sulphur increases the risk of high temperature cracking, and it is preferable to employ high ferrite 18 % Cr/10 % Ni or 29 % Cr/9 % Ni type filler metals.

– Welding stainless or low alloy martensitic and ferritic steels together or to austenitic stainless steels, apart from steels having a high sulphur content as mentioned above.

– Welding assemblies for very low temperature service. This category includes 9 % nickel martensitic steels for use at down to −196 °C. In §2.2 it was pointed out that these steels can be welded using an 18 % Cr/10 % Ni filler metal with an adequate ferrite content. In general terms, it is possible to employ any austenitic filler metal ensuring good weldability and high ductility at low temperature; however, it is often necessary that the weld should have higher mechanical characteristics than those obtained with conventional 18 % Cr/10 % Ni filler metals giving a ferrite content below 8 %. These requirements are fulfilled by nickel-base filler metals containing chromium, iron and some hardening additions of molybdenum, tungsten, niobium, as exemplified by the type considered in chapter 7, §4.3. Fig. 9.8 represents the fusion boundary of a joint in 9 % nickel steel welded with this type of filler metal, in which even the most diluted weld metal contains no visible martensite. This is not the only solution. Less alloyed fully austenitic filler metals have been used [8] containing 17 % chromium, 14 % nickel, a high quantity of manganese to avoid cracking, and hardened by substantial additions of tungsten and carbon. These have enabled 9 % nickel steels to be welded under good conditions.

This brief survey of the possibilities offered by the nickel–chromium–iron alloys shows that they are universal filler metals, which afford a solution to almost all problems arising in heterogeneous welds. Unfortunately, they are more expensive than the conventional stainless steel filler metals, and their use is therefore limited, in spite of their technical advantages.

References

[1] H. Granjon. *Rev. Mét.* **51** (1954) 221.
[2] A. Portevin. Lettre à l'éditeur. *Metal Progress*, **33** (1938) 169.
[3] D. Séférian and J. Vallée. *Journée d'Étude d'Équipement pétrolier*, Liège, 1956.
[4] H. Thielsh. *Weld. J. Res. Sup.* **31** (1952) 37 s.
[5] D. Slater and J. M. Winn. Second Commonwealth Welding Conf., London (1965) Inst. Weld.
[6] J. F. Eckel. *Weld. J. Res. Sup.* **43** (1964) 171 s.
[7] J. T. Tucker and F. Eberle. *Weld. J. Res. Sup.* **35** (1956) 529 s.
[8] P. Amiot, M. Hollard, J. Hochmann, R. Quigna and J. Rosenzweig. *Rev. Mét.* **62** (1965) 835.

10

APPLICATION OF WELDABILITY TESTS TO STAINLESS AND HEAT-RESISTING STEELS

1 General comments

The necessity for carrying out tests on welds results from the following considerations, which are inherent in the fabrication of welded constructions:

– It must be possible to make the welded joints.
– The welds must withstand service stresses and aggressive environments.
– The construction must satisfy particular specifications, or conditions defined by codes.

In the first instance, one can consider tests performed by the manufacturer principally to qualify the welding procedure or to control the weld quality. In this type of test, the joint geometry, parent material and filler metal are already defined, and it is known that welds can be made possessing the desired properties.

In tests aimed at qualification of the welding procedure, welds similar to the real ones are carried out, and the process and welding conditions are selected so that the joints satisfy the conditions imposed by the relevant specification or code. The production welds are inspected by non-destructive or destructive tests performed on extension plates welded at the same time as the final assembly. These tests are intended to establish the weld quality in accordance with the given specification or code.

These tests are carried out subsequent to more general ones which the metallurgist must perform to study the physical possibility of accomplishing a weld, and the suitability of this weld for use under service conditions. They are also needed for alloy development so that the welds may better fulfil service requirements. In the following, these tests will be termed 'weldability tests', for the sole purpose of distinguishing them from the previous ones and will be described briefly in this chapter. It should be noted that weldability tests are applicable to all categories of steels, and that, in most cases, they have been designed for ordinary or low alloy steels, and subsequently applied to stainless and heat-resisting steels.

2 Weldability tests

The concept of weldability is complex, and for simplification, three types of weldability may be distinguished [1]:

– *Operative weldability*, considering the feasibility of obtaining a welded joint between two similar or different metals, as well as the practical conditions under which this welded joint can be achieved.

– *Local weldability*, based on the study of structural or compositional changes caused by a welding operation in the different regions of the weld, together with the effects arising from these individual changes upon the physical and chemical properties of the weld.

– *Overall weldability*, concerning the ability of the weld to withstand successfully the various types of loading (including residual welding stresses) to which it will be subjected in service.

Most weldability tests consist of making welds in testpieces designed to evaluate a particular type of weldability, generally under more severe conditions than normally encountered in service. In fact, the majority of tests examine, to some extent, all three types of weldability given above, as these facets of weldability are not independent.

Four types of tests are differentiated below, which are generally performed in the metallurgical study of the weldability of stainless steels, but which are not specific to these materials:

- Fusion and weld metal soundness tests.
- Weld cracking tests.
- Tests related to the applied properties of the welds.
- Examination of the physico-chemical structure. Such examination can, in fact, be advantageously carried out when performing a test of any of the three previous types.

3 Tests for fusion and weld metal soundness

These tests can be as follows:

- Melt runs, or deposits carried out on a metal plate.
- Welds made under low restraint conditions.
- Thick deposits for weld metal analysis.

They are intended to demonstrate:

– That a regular weld bead can be made, which wets the base metal properly, and whether or not refractory oxides are formed which may hinder the fusion of the following runs.

– Compositional changes of the weld deposit as compared to the filler metal, resulting from the oxidation of reactive elements (chromium, titanium, etc.). These tests often constitute the first criterion for the selection of the welding process, to ensure that weld deposits of the desired composition can be obtained.

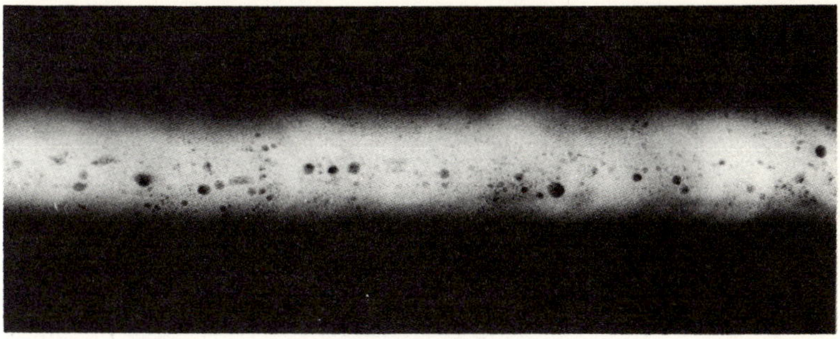

Fig. 10.1 Radiograph of a porous weld in a nickel-base alloy.

– The risk of weld metal porosity as a function of the welding process employed, of the joint configuration, of impurities in the consumable and of the measures taken to protect the molten pool against the atmosphere. The porosity can be detected by examining sections through the weld metal (fig. 3.3), fracture surfaces or radiographs. As an example, fig. 10.1 shows a radiograph of a nickel-base alloy weld containing porosity. The results obtained also give a guide to the welding process selection, or to suitable compositional changes such as the addition of titanium to the filler metal of fig. 10.1.

4 Cracking tests

There are numerous types of cracking tests, as well-indicated by document IIS 93–62 of the International Institute of Welding. In general, each test was designed to give an answer to a particular problem, but all of them can, if necessary, be applied to the various categories of stainless steels. In the following, an attempt is made to classify them into three groups, giving a few examples.

4.1 Overall tests

These welding tests are carried out on plates generally designed to put the weld under severe restraint conditions in which cracking is likely to occur. The different tests can be distinguished by the degree of restraint which they involve, but this concept is purely qualitative.

All tests with a high degree of restraint are applicable to austenitic stainless steels. The RD test[2], shown schematically in fig. 10.2, can be considered as an example. In this test the restraint is achieved by the fact that the test plates are pressed against one another, while the assembly is prevented from bending by the form and rigidity of the fixture on which the plates are mounted.

Another very severe test consists of machining a circular groove in a very thick plate, and filling the groove with weld metal.

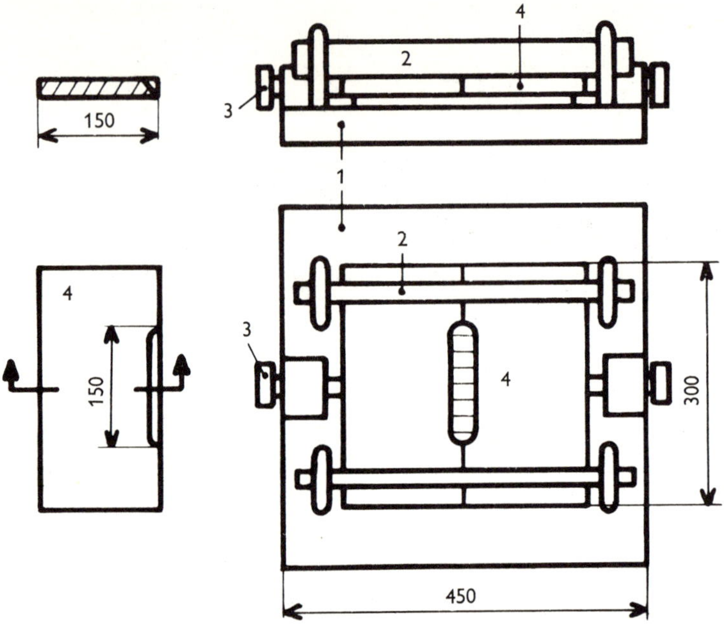

Fig. 10.2. RD test. 1, support plate; 2, cross bars; 3, lateral tightening; 4, test plate.

In these tests, the degree of restraint can be adjusted qualitatively by varying the thickness of the test plates.

There are analogous tests applicable to thin plates. The AIR 9113 specification describes a butt weld test, in which the plates are highly restrained in a massive fixture.

In the case of martensitic stainless steels, the restraint does not play an exclusive role in crack formation, and the test must be accompanied by preheating and post-heating.

It is, of course, possible to vary the thermal conditions under which the above-mentioned tests are carried out, but this is not always easy because of the size of the fixtures; moreover, these additional parameters multiply the number of tests to be performed, and consequently it is preferable to use the simplest test possible.

The 'controlled thermal severity' test (CTS test of English origin) is an example of a test which is simple to carry out and is applicable to martensitic steels [3]. It is composed of two plates, with varying thickness, fastened to one another by a central bolt and two lateral weld beads called 'anchor runs', as shown in fig. 10.3. The test beads C_2 and C_3 are carried out in this order, the assembly being preheated to the selected temperature. It is very easy to vary the cooling rate in the critical zone by modifying the plate thicknesses and the preheating temperature. Further, as the assembly is not symmetrical, the bead C_3 cools

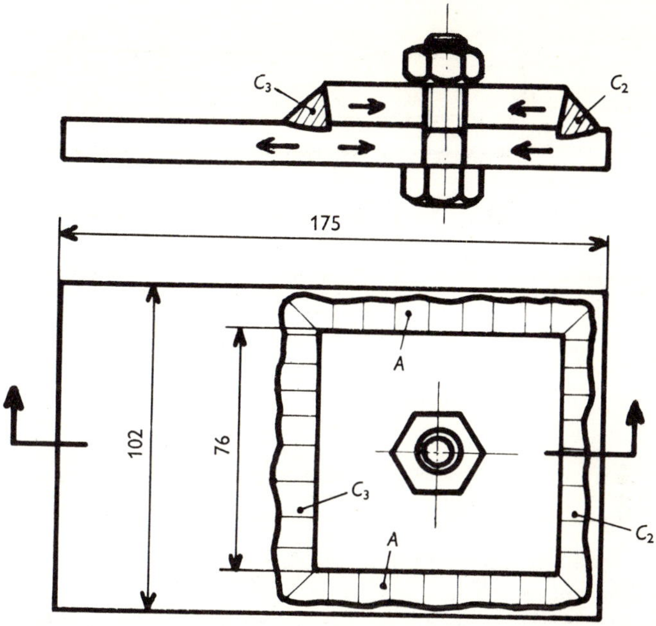

Fig. 10.3. CTS test. *A*, anchor runs; C_2, bithermal bead; C_3, trithermal bead.

more rapidly than the bead C_2 (its 'thermal severity' is higher). In effect, heat flow from bead C_2 takes place by two equivalent paths, symbolised by arrows on the section of fig. 10.3, whereas there are three equivalent paths in the case of bead C_3. For this reason, C_2 and C_3 are respectively termed the 'bithermal bead' and the 'trithermal bead'. This test can, of course, be applied to austenitic steels, but the thermal severity concept is not particularly significant with these alloys.

4.2 Overall tests with variable degree of restraint

When the preceding tests are applied to austenitic stainless steels, it is very difficult, if not impossible, to evaluate the cracking tendency. It is generally feasible, by always using the same test, to classify different materials, but for practical purposes, this classification has only two identification categories, namely 'good' behaviour (no cracking) and 'bad' behaviour (with cracks).

With these tests, it is impossible to evaluate the cracking tendency so that comparison can be made with practical behaviour. This often leads to the selection of a very severe test, sometimes bearing no relation to the restraint usually encountered in welded constructions, in order that the most conservative, and therefore safe, estimate of crack susceptibility can be obtained.

In consequence, some investigators have developed tests with a variable

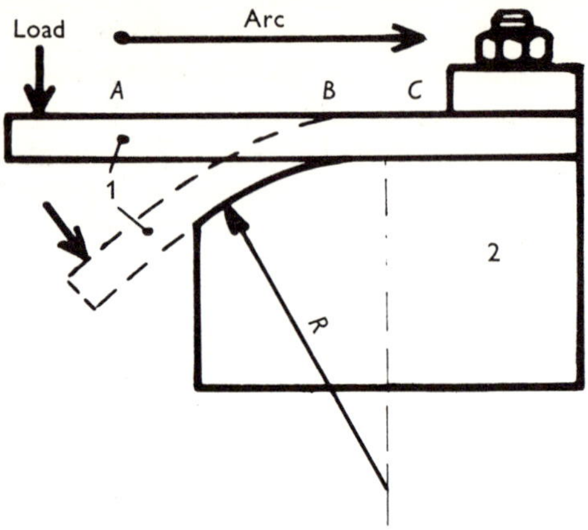

Fig. 10.4. Test with variable degree of restraint (Varestraint test). 1, test plate; 2, supporting block; *ABC*, arc trajectory.

degree of restraint, imposed by some factor independent of the thickness of the welded plate.

In some instances, a weld is performed while the edges of the groove are pulled apart. Such is the case for a test in which a fillet weld is carried out while the fillet angle opens progressively; this is also the case [4] when a weld bead is deposited in the groove formed by two cylinders placed parallel to each other and turning in such a way that the weld is under strain. In both these tests, the cracking tendency is estimated by the angular rate of the groove opening above which cracks are formed in the weld metal.

Based on the same idea, the test shown schematically in fig. 10.4 has been recently proposed [5], and is attractive by virtue of its simplicity. The testpiece is a flat plate 1 fixed at its extremity on a rigid block 2, the surface of which comprises part of a cylinder of radius *R*. An argon TIG melt run is made from point *A* to point *C*. When the arc reaches the fixed point *B*, the plate is rapidly bent against the cylindrical part of the block. The test severity is expressed in terms of the strain on the outer surface of the plate, the strain varying with the plate thickness or with the radius *R*. It is relatively easy to evaluate the susceptibility to cracking by the strain above which cracks appear, or by the total length of the cracks for a given strain, as all the cracks formed in the weld metal or in the HAZ appear at the surface. This test has the further advantage of not requiring the fabrication of a filler metal, but requires only plates of the given composition of interest. In this form, however, it obviously cannot be applied to heterogeneous welds.

The S.N.E.C.M.A. variable restraint test finds application for thin sheet material [6]. This consists in carrying out a circular fusion run on a plate fixed at its periphery and deformed by a central punch. The degree of

restraint is determined by the amplitude of the imposed deforma
it is relatively easy to evaluate the extent of cracking by the len
cracks visible on the surface.

4.3 Local weldability testing

In the tests described above, an actual weld is made, and all the three aspects of weldability mentioned in §2 are involved to some extent. The local weldability test, on the other hand, is meant to study, in isolation, the effect of the welding thermal cycle upon the mechanical strength and ductility of the metal, measured at the temperatures over which shrinkage stresses appear. These are, in fact, the characteristics that ultimately determine the resistance to hot cracking [7-9].

This test employs a tensile specimen which is generally resistance heated using a programming device to reproduce the thermal cycle experienced in various regions of the HAZ. At a certain point of the cycle, the specimen is fractured by tensile strain applied at an adjustable speed, and the tensile strength and ductility of the metal are measured.

The test usually involves heating to a temperature close to the melting point, with strain being applied on the cooling part of the selected thermal cycle.

This test gives an indication of the loss in high temperature ductility or mechanical strength that can result from a rapid heating of the metal to near the solidus. Thus, the capacity of the material either to resist shrinkage stresses or to accommodate them by plastic deformation can be assessed.

A variant of this test involves the use of a specimen placed in a refractory tube, in such a way that the metal can be melted and the tensile test performed during the cooling after solidification. This enables the approach to be extended to study weld metal behaviour.

4.4 Crack detection

Cracks can be detected by visual examination when they appear at the surface. This method is often sufficient in the case of martensitic steels, as the cold cracking which can occur in these steels generally propagates over a considerable distance.

In austenitic stainless steels, the cracks can be too fine to be detected visually, and a penetrant test may therefore be carried out on successive runs. However, this allows detection only of cracks appearing at the surface, and austenitic stainless steels may be sensitive to the formation of fine internal cracking. These cracks can be revealed only by examining sections taken from the weld, with a microscope if necessary, but it may then be illusionary to try to evaluate the cracking susceptibility from the number or total length of the cracks.

In the case of tests intended for quality control of welded constructions, or for qualification of the welding procedure, ultrasonic or radiographic examination is often employed to detect defects present in the welds, with particular reference to cracking. These methods are useful for weldability tests on martensitic stainless steels in which cracks that have not caused fracture are nevertheless relatively important. The detection of the small cracks that can be found in austenitic steels is easy only when they are favourably orientated with respect to the X-ray or ultrasonic beam waves, although the use of finely focused ultrasonic techniques can reveal the presence of very small defects. It should, however, be noted that the columnar structure of fully austenitic weld metals, which is often rather coarse, can make the welds relatively opaque to ultrasonic waves.

5 Tests related to the service properties of welded joints

The purpose of these tests is to verify that the properties of the weld permit its use under service conditions similar to those of the base metal. Tests carried out upon stainless and heat-resisting steels in this context are mainly mechanical and corrosion resistance tests.

5.1 Mechanical tests

These can be considered in two ways, depending on whether the principal interest lies in the local characteristics of the various regions of the weld, or in the overall characteristics of the welded joint.

The properties of the weld metal can be determined by preparing a deposit thick enough for conventional or reduced-size all-weld metal specimens to be taken. Normal tensile or impact tests, creep tests and, more rarely, fatigue tests can be performed in this way. Tensile specimens are usually taken in the direction of the weld, although this gives only partial information about the weld metal behaviour, because of the anisotropy of solidification. Since the notch dimensions are relatively small, impact testing can enable assessment to be made of the weld metal properties in different directions. Impact specimens can consist totally of weld metal, or be taken across a welded joint; in the latter case, it is possible to site the notch position in such a way that only a particular region of the weld metal or HAZ is tested, which is impossible with conventional tensile specimens. In the same way, hardness tests carried out on a weld cross-section allow complete examination of the local characteristics of the different regions constituting the weld.

The evaluation of weld metal strength by tensile tests on all weld metal specimens gives only an incomplete idea of the performance to be expected from a welded joint. Indeed, from the mechanical point of view, a weld constitutes a composite assembly in which local properties do not determine service behaviour independently of the properties of other regions, and the fact that the weld metal or HAZ has a tensile strength inferior to that of the base metal does not imply that the tensile strength of the welded joint is that of the softest region. In fact, the hard regions adjacent to the softest region prevent its deformation and necking down. This may be the case, for instance, in a circumferentially welded tube operating at the bursting stress. Because of the distribution of stresses in the metal, fracture always occurs parallel to the tube axis. Even when the weld is weaker than the base metal, bursting always involves the deformation and rupture of unaffected metal, which therefore contributes to the overall strength of the assembly.

Moreover, in austenitic steels working at normal ambient temperature, the presence of parent material which has not been affected by the welding cycle promotes preferential work-hardening of any soft regions when the weldment is put into service, or during a tensile test.

It is therefore advisable to measure the overall strength of the welded joint whenever possible, as this can be considerably higher than the lowest local strength, and because, ultimately it is the only strength of practical importance. Evaluation of overall strength can be carried out on specimens containing the weld and unaffected base metal, taken in different orientations with respect to the weld.

Overall toughness tests, such as Robertson or Pellini tests, can be applied to martensitic stainless steels and, if really necessary, to ferritic steels. They are not, of course, applicable to austenitic stainless steels, in which welding does not cause embrittlement.

5.2 Corrosion tests

It has been pointed out that the welding operation can produce a local sensitivity to intercrystalline corrosion in ferritic stainless steels, certain austenitic stainless steels and nickel-base alloys. In practice, it is only on austenitic steels, and, to a lesser extent, nickel alloys that the tests for sensitisation by welding are performed.

These corrosion tests can take numerous forms, but all consist of making a weld in relatively thin material, and immersing it in a reagent defined by certain specifications or standards for a fixed period. As an example, the procedure selected for plate products by l'Etablissement d'Indret de la Marine (Instruction 47) to reveal sensitisation to intercrystalline corrosion by carbide precipitation in austenitic stainless steels is shown in fig. 10.5. It consists of welding two plates of about 50 mm × 100 mm (2 in. × 4 in.) together, and then depositing a bead on the assembly perpendicular to this weld. A specimen is taken from the weld and immersed in a copper sulphate solution which attacks the base metal zones sensitised by the weld, and the areas of base metal and weld sensitised by the second bead. The extent of corrosion is usually revealed qualitatively by a bend test which causes the appearance of fissures in the corroded regions, or by microscopic examination of the depth of corrosion.

When it is desired to test the resistance to intergranular corrosion of the

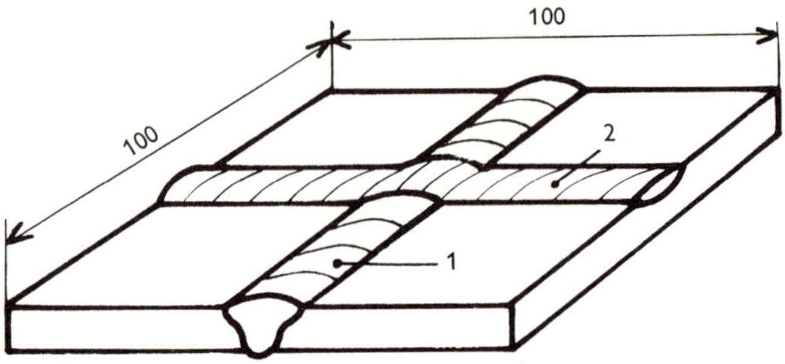

Fig. 10.5. Test for sensitisation to intercrystalline attack on welding. 1, weld; 2, cross bead.

weld metal alone, a large weld pad is made and a section taken. The sample is then subjected to attack in the reagent, either in the as-welded condition, or after furnace sensitisation similar to that used for the parent material.

It should be appreciated that the conventional intergranular corrosion test applied to unwelded material, which consists of immersing a sample sensitised by furnace heat treatment in the test solution, is actually a test of local weldability. This procedure is aimed principally at evaluating the ability of the steel to tolerate welding and hot-working operations without any risk of intergranular corrosion in service, when there is no stabilising treatment after such operations.

Certain tests for resistance to intercrystalline corrosion cause such a severe attack that individual grains may be dislodged from the metal, with a consequent high loss in specimen weight. This is the case with the Huey test in hot nitric acid, or with some tests applied to nickel-base alloys. In this situation, the estimation of the weight loss can constitute the result of the test, when unwelded material is being examined. When the test is applied to a welded joint, the localised nature of the sensitisation due to welding means that a result expressed in terms of overall loss in weight has no significance. If an evaluation of the corrosion is desired, it is necessary to measure the depth of the attack by microscopic examination.

These comments regarding intercrystalline corrosion are also applicable to other tests intended to show the overall effect of the welding operation upon the corrosion resistance of a joint: the test can be performed by immersing the welded joint in a suitable aggressive medium, and then examining the form and intensity of the attack at the surface of the specimen or on sections. This principle can also be applied to tests of oxidation and corrosion behaviour at high temperature.

6 Metallurgical examination

In the study of the weldability of stainless and heat-resisting steels, all the tests described above can be used for metallographic examination to determine the nature of the physico-chemical constituents and the various types of heterogeneity caused by the welding cycle.

Sections taken from different regions of the weld may be examined by microscopic techniques to determine the nature and distribution of the phases present in the weld, by local microanalysis to determine the chemical composition of the various phases and the extent of any segregation, and also by microhardness tests.

In the previous chapters, the aim has been to show the relationship existing between the physico-chemical structure of a weld, and its ability to withstand the shrinkage stresses, mechanical loading and chemical environments to which it is subjected in service. Thus, the importance of metallographical examination will not be pursued. It will suffice to point out that the experience of a metallurgist quite often enables him to predict the properties of a weld from the study of its physico-chemical structure.

References

[1] H. Granjon and A. Audigé. *Rev. Mét.* **58** (1961) 330.
[2] A. J. Elliot and W. I. Pumfrey. *J. Iron Steel Inst.* **163** (1949) 369.
[3] H. Granjon. *Rev. Mét.* **57** (1960) 319.
[4] H. Fredriks and F. F. Ates. *Br. Weld. J.* **10** (1963) 620.
[5] W. F. Savage and C. D. Lundin. *Weld. J. Res. Sup.* **44** (1965) 433 s.
[6] B. Blanchet. *Soudage Techniques connexes*, **16** (1962) 18.
[7] E. F. Nippes, W. F. Savage and G. Grotke. *Weld. Res. Counc. Bull. Ser.* (1957) no. 33.
[8] C. S. Williams. *Weld. J. Res. Sup.* **42** (1963) 1 s.
[9] C. H. Kreischer. *Weld. J. Res. Sup.* **42** (1963) 49 s.

INDEX